SpringerBriefs in Applied Sciences and Technology

SpringerBriefs present concise summaries of cutting-edge research and practical applications across a wide spectrum of fields. Featuring compact volumes of 50 to 125 pages, the series covers a range of content from professional to academic.

Typical publications can be:

- A timely report of state-of-the art methods
- An introduction to or a manual for the application of mathematical or computer techniques
- A bridge between new research results, as published in journal articles
- A snapshot of a hot or emerging topic
- An in-depth case study
- A presentation of core concepts that students must understand in order to make independent contributions

SpringerBriefs are characterized by fast, global electronic dissemination, standard publishing contracts, standardized manuscript preparation and formatting guidelines, and expedited production schedules.

On the one hand, **SpringerBriefs in Applied Sciences and Technology** are devoted to the publication of fundamentals and applications within the different classical engineering disciplines as well as in interdisciplinary fields that recently emerged between these areas. On the other hand, as the boundary separating fundamental research and applied technology is more and more dissolving, this series is particularly open to trans-disciplinary topics between fundamental science and engineering.

Indexed by EI-Compendex, SCOPUS and Springerlink.

Najwa Syakirah Hamizan · Solehuddin Shuib ·
Amir Radzi Ab Ghani

Hip Prosthesis

CAD Modeling, Finite Element Analysis
(FEA) and Compressive Load Testing

Najwa Syakirah Hamizan
School of Mechanical Engineering
Universiti Teknologi MARA (UiTM)
Shah Alam, Selangor, Malaysia

Solehuddin Shuib
School of Mechanical Engineering
Universiti Teknologi MARA (UiTM)
Shah Alam, Selangor, Malaysia

Amir Radzi Ab Ghani
School of Mechanical Engineering
Universiti Teknologi MARA (UiTM)
Shah Alam, Selangor, Malaysia

ISSN 2191-530X   ISSN 2191-5318 (electronic)
SpringerBriefs in Applied Sciences and Technology
ISBN 978-981-96-1469-1   ISBN 978-981-96-1470-7 (eBook)
https://doi.org/10.1007/978-981-96-1470-7

This Springer imprint is published by the registered company Springer Nature Singapore Pte Ltd.
The registered company address is: 152 Beach Road, #21-01/04 Gateway East, Singapore 189721, Singapore

If disposing of this product, please recycle the paper.

Preface

Introduction

The book is motivated by the challenge of achieving optimal performance in hip implants, especially considering the diverse mechanical loads and conditions they endure in the human body. The research focuses on improving the design and finite element analysis (FEA) of hip implants to enhance their longevity and functionality, aiming to provide better outcomes for patients undergoing hip replacement surgery. This motivation is driven by the need for implants that can withstand complex stress patterns and minimise complications, ultimately improving patient quality of life.

Share Any Personal Experiences

During the early 90s, I had the privilege of meeting two renowned orthopaedic surgeons, Dr. Mazwar and Dr. Roslan, in Toledo, Ohio, where they were on a medical attachment. These interactions were more than just professional exchanges; they were pivotal moments that shaped my career path. Dr. Mazwar and Dr. Roslan, both leading figures in Malaysia's orthopaedic community, shared their deep knowledge and passion for advancing medical technology, particularly in hip prosthesis. Their dedication to improving patient outcomes and their innovative approaches to surgery inspired me to delve into research focused on hip prostheses. Their influence was instrumental in my decision to contribute to this field through the book I'm writing, where I hope to advance the understanding and development of hip implants using modern technologies like AM 3D printing. Their legacy continues to inspire my work, driving me to push the boundaries of what is possible in orthopaedic research and design.

Purpose

This book is important because it addresses the critical challenge of improving the design and performance of hip implants, which are essential for enhancing patient outcomes in hip replacement surgeries. The book focuses on utilising finite element analysis (FEA) to optimise implant designs, ensuring they can withstand complex mechanical loads and reduce the risk of failure. It is hoped that readers will gain a deeper understanding of the significance of implant design and the role of FEA in predicting and enhancing implant performance, ultimately leading to better, more durable solutions in orthopaedic surgeries.

Acknowledgments

We would like to extend our appreciation to Prof. Amran Ahmed Shokri, Orthopaedic Surgeon, USM, for his invaluable guidance, expertise and contributions, particularly in providing medical insights into the hip implant. We are also deeply grateful to Dr. Izhar Aziz and 3D Gens Sdn. Bhd. for their unwavering support and collaboration. The success of this research would not have been possible without their commitment to fostering innovation and bridging the gap between academia and industry.

We would like to express our sincere gratitude to the Research Management Centre (RMC), for their generous funding and support namely UiTM (600-IRMI/DANA 5/3/BESTARI (0004/2016)), 600-RMC/KEPU 5/3 (009/2021), and 600-IRMI/MYRA 5/3/MITRA (001/2017)-1. This work would not have been possible without their assistance.

Our appreciation also goes to all the staff who provided the facilities and assistance during the testing and manufacturing processes. Their cooperation and willingness to assist were fundamental to the success of this study. A special thanks to our colleagues and friends who offered valuable insights, engaging discussions, and helpful suggestions that have enriched this book.

Lastly, we sincerely thank our family for their unwavering support and encouragement throughout the preparation of this book.

Outline

This book is structured into six significant chapters, each contributing to a comprehensive understanding of the problem being studied, ultimately aiding in the development of a hip prosthesis using AM 3D printing technology. Chapter 1, "Design and Development of Hip Implant Using Finite Element Analysis (FEA)," introduces the research background, detailing the problem statements, research questions and objectives. It also defines the scope and current issues of hip implants, ensuring the research remains focused and timely. Chapters 2 and 3, "Anatomy of Hip and Bioma-terial for Hip Implant" and "Pugh Chart Method and Finite Element Analysis (FEA) of Hip Implant," review relevant literature, covering topics such as hip anatomy, total hip arthroplasty (THA), FEA, jig preparation and testing of fabricated hip prostheses. These reviews are essential for comparing past findings with new research outcomes. Chapter 4, "Development of Hip Implant," describes the phases of designing and analysing the new hip implant using CATIA and ANSYS software, including jig preparation and compressive load testing to validate the design. Chapter 5, "Finite Element Analysis (FEA) of Modified Design of Hip Implant," presents the find-ings from FEA, comparing the hip implant with and without PMMA through both simulations and compressive load testing and validating the data against previous research. The final chapter, (Chap. 6) "Fundamental of Design and Development of Hip Implant," summarises the research process and offers recommendations for future advancements in this field. Visual aids such as figures, tables and graphs are utilised throughout to enhance understanding of the outcomes.

Conclusion

The exploration of hip implant and the application of finite element analysis (FEA) have not only been academically challenging but also deeply rewarding. Throughout this process, I have grown both as a researcher and as an individual, gaining insights into the intricate balance of biomechanics and engineering. Finally, we dedicate this work to all the patients who inspire innovation in the field of orthopaedic surgery. May this research contribute, even in a small way, to enhancing their quality of life.

Thank you.

Shah Alam, Malaysia   Prof. Ir. Ts. Dr. Solehuddin Shuib
                                                        Najwa Syakirah Hamizan
                                            Assoc. Prof. Dr. Amir Radzi Ab Ghani

About This Book

In the rapidly evolving field of orthopaedic implants, 3D printing technology has brought forth remarkable advancements, yet its full potential remains untapped, particularly in the development of 3D-printed hip prostheses. While the acetabular cup has seen successful integration into clinical practice, the femoral stem lags, with no documented cases of its use in patients. This book delves into this gap, presenting ground-breaking research that redefines the design and analysis of hip prostheses.

Through a meticulous evaluation of the existing Charnley hip prosthesis, the author employs the Pugh chart method, advanced CAD modelling and finite element analysis (FEA) to explore innovative modifications aimed at minimising stress shielding and optimising weight. The book also highlights the challenges and successes of combining FEA with experimental testing, offering a rare comparison between simulated and real-world performance.

Featuring custom-designed jigs, compressive load testing and the use of cutting-edge materials like titanium alloys, this work sets the stage for future innovations in hip implant design. Whether if you are a researcher, clinician or engineer, this book provides a comprehensive guide to the future of hip implant technology, where design meets experimental validation.

Contents

About the Authors

Najwa Syakirah Hamizan obtained her Bachelor of Engineering (Honours) in Mechanical Engineering from Universiti Teknologi MARA (UiTM), Malaysia, in 2020. She completed her Master of Science in Mechanical Engineering from UiTM in 2024, with a thesis focused on the design and evaluation of a customised hip prosthesis. Throughout her MSc studies, she served as a UiTM Postgraduate Teaching Assistant (UPTA), assisting in laboratory works for mechanics of materials, thermal engineering, fluid mechanics and finite element method and analysis courses for undergraduate students. She is a registered graduate engineer with the Board of Engineers Malaysia (BEM) and a graduate member of the Institution of Engineers Malaysia (IEM).

Ir. Ts. Dr. Solehuddin Shuib is a professor of biomechanics at the School of Mechanical Engineering, Universiti Teknologi MARA, Malaysia. He holds a BS in Mechanical Engineering from the University of Alabama at Birmingham, Alabama, USA, a master's in mechanical engineering from the University of Toledo, Ohio, USA, and a Ph.D. in Mechanical Engineering from Universiti Putra Malaysia, Malaysia. He has authored or co-authored more than 70 refereed journal publications and presented at over 100 technical meetings. His research interests are in biomechanical engineering and medical devices.

Ir. Dr. Amir Radzi Ab Ghani is an associate professor at the School of Mechanical Engineering, Universiti Teknologi MARA, Malaysia. He graduated from the University of Liverpool, 1994 in Mechanical Engineering. In 1995, he obtained his master's from the same university. He completed his Ph.D. at UiTM in 2014, specialising in the impact performance of metallic structures for automotive applications. Ir Dr Amir Radzi has supervised many post-graduate students and published more than 50 papers in international journals and conference proceedings. His current interests are structural impact and crashworthiness, and automotive engineering. His knowledge and expertise in structural analysis enable him to be involved in projects of various areas such as electric mobility, oil and gas, sea transportation and biomechanics.

Abbreviations

3D	Three-Dimensional
ABC	Acrylic Bone Cement
AJRR	American Joint Replacement Registry
AM	Additive Manufacturing
AOANJRR	Australian Orthopaedic Association National Joint Replacement Registry
ASTM	American Society for Testing and Materials
CAD	Computer-Aided Design
CJRR	Canadian Joint Replacement Registry
CNC	Computer Numerical Control
CPC	Calcium Phosphate Bone Cement
DHR	Danish Hip Arthroplasty Register
DIC	Digital Image Correlation
DMLS	Direct Metal Laser Sintering
FDM	Fused Deposition Modelling
FE	Finite Element
FEA	Finite Element Analysis
FEM	Finite Element Method
FOS	Factor Of Safety
GHV	Gentamicin High Viscosity
GPC	Glass Polyalkenoate Cement
HV	High Viscosity
IR 4.0	Industrial Revolution 4.0
ISO	International Organization for Standardization
JSRA	Japanese Society for Replacement Arthroplasty
LROI	*Landelijke Registratie Orthopedische Implantaten* or Dutch Arthroplasty Register
MV	Medium Viscosity
NAR	Norwegian Arthroplasty Register
NJR	National Joint Registry
NORM	National Orthopaedic Registry Malaysia

NZJR	New Zealand Joint Registry
PAA	Polyacrylic Acid
PE	Polyethylene
PHS	Public Health Scotland
PLA	Polylactic Acid
PMMA	Polymethyl Methacrylate
SAR	Swedish Arthroplasty Registry
SDG	Sustainable Development Goals
SLM	Selective Laser Melting
THA	Total Hip Arthroplasty
THR	Total Hip Replacement
TPMS	Triply Periodic Minimal Surfaces
UHMWPE	Ultra-High Molecular Weight Polyethylene
UTM	Universal Testing Machine

Nomenclatures

cm^3	Centimetre cubic (unit)
CoCr	Cobalt-Chromium alloy
CoCrMo	Cobalt-Chromium-Molybdenum alloy
CoNiCrMo	Cobalt-Nickel-Chromium-Molybdenum alloy
COVID-19	Coronavirus 2019
GPa	Giga Pascal (unit)
g	Gram (unit)
hr	Hour (unit)
kg	Kilogram (unit)
kN	Kilo Newton (unit)
MYR	Malaysian Ringgit (currency)
μm	Micrometre (unit)
mm	Millimetre (unit)
mm^3	Millimetre cubic (unit)
MPa	Mega Pascal (unit)
Ti-6Al-4V	Titanium alloy (90% Titanium, 6% Aluminium and 4% Vanadium)
USD	United States Dollar (currency)

Symbols

&	And
F	Applied Force
ΔL	Amount of Deformation or Change in Length
Co	Cobalt
A	Cross-Sectional Area
°C	Degree Celsius
Mg	Magnesium
N/A	Not applicable
Ni	Nickel
L_0	Original or Initial Length
%	Percentage
$\pm$	Plus-or-Minus Sign
ν	Poisson's Ratio
σ	Principal Stresses
RM	Ringgit Malaysia
τ	Shear Stress
Ti	Titanium
σ_{vm}	Von Mises Stress
E	Young's Modulus
US $	United States Dollar

List of Figures

List of Tables

Chapter 1
Design and Development of Hip Implant Using Finite Element Analysis (FEA)

1.1 Introduction

This chapter provides the research background, main challenges, research questions and objectives of the study. In addition to that, the scope and current issues that are faced by the product due to the constraints and limitations are mentioned to ensure that the research carried out is within the field of interest and is conducted within the given time frame. The significances of the study are also highlighted in this section.

1.2 Research Background

Following the knee joint, the hip joint is the second largest weight-bearing joint of the human body [1]. It links the pelvis to the femur, commonly known as the thigh bone. As it carries the weight of the torso or the upper body, it is exposed to high-stress daily movement from activities such as walking, jumping, running and climbing, which can threaten its function ability over time as one gets older [2]. Conditions like various types of arthritis, infections and other causes of joint deformities or trauma-induced damage appear to aggravate and worsen the hip joint's condition instead of promoting healing. The progressive joint anomalies may lead to extreme discomfort, non-functional ambulatory status and claudication (muscle pain, cramps or fatigue that occurs during physical activity but are relieved when one rests). These symptoms are prone to exhibit more in the hip joint rather than in the knee or ankle joints [3].

Total hip arthroplasyt (THA), which is also referred to as total hip replacement (THR), is one of the highly successful procedures for joint restoration, along with surgeries involving knee, ankle, elbow and shoulder joints. This surgical intervention involves the implantation of an artificial joint called a hip prosthesis, into the femur, intended for it to perform similar purposes as the actual joint [2].

© The Author(s), under exclusive license to Springer Nature Singapore Pte Ltd. 2025
N. S. Hamizan et al., *Hip Prosthesis*,
SpringerBriefs in Applied Sciences and Technology,
https://doi.org/10.1007/978-981-96-1470-7_1

Throughout the years, a significant number of THA surgeries have been conducted in various countries. In the UK alone, data collected from the National Joint Registry (NJR) since it started to collect information in the year 2003 reveals that there were over 1,344,000 hip replacement surgeries that have taken place [4]. Similarly, in the USA, over 1,060,000 surgeries have been performed since 2012, according to the data obtained from the American Joint Replacement Registry (AJRR) [5]. Australia has had nearly 800,000 surgeries executed since 1999, as documented by the Australian Orthopaedic Association National Joint Replacement Registry (AOANJRR) [6].

To gain further insight into THA, additional information gathered from the respective 2022 annual reports for THA surgeries carried out in various countries that can be found are compiled and summarised in Table 1.1, to better understand the impact of THA worldwide. The number of THA surgeries globally has been steadily increasing, although most countries experienced a drop in THA cases, particularly during the years 2020 and 2021 due to the impact of the 2019 pandemic coronavirus (COVID-19), which had a profound impact on the healthcare systems worldwide [4–8]. Despite the setback, the overall trend quickly increases as COVID-19 slowly loosens its pandemic-related restrictions and enters the endemic phase transition.

Nonetheless, the process of obtaining data on the statistics of THA surgeries performed in Asia has been extremely challenging [19]. This difficulty may be attributed due to the lack of standardised, digitalised, synchronised and uniform reporting systems across countries within the continent [20]. In the case of Malaysia, for example, while there have been some sporadic reports available from different hospitals, the absence of a cohesive and uniform reporting system has made it impossible to effectively track THA surgeries conducted within the country. The National Orthopaedic Registry Malaysia (NORM) had previously released an annual report back in 2009, and that too only contained information on hip fractures that happened in the country [21]. However, no further report has been made available in the public domain following the initial release of the publication [22]. This situation creates a difficult circumstance surrounding the data acquisition on THA surgeries in Malaysia, proving the lack of a standardised reporting system.

1.3 Main Challenge

The annual increase in hip replacements has led to a higher demand for hip implants that can last for an extended period [23]. However, the traditional manufacturing methods, such as Computer Numerical Control (CNC), casting and machining, used to design hip prosthesis are time-consuming and often require additional post-processing, contributing to lengthy production times and high costs. At present, Malaysia relies on imported hip prosthesis, resulting in extended waiting periods due to shipping times. In line with the Industrial Revolution 4.0 (IR 4.0), additive manufacturing (AM) technology has become one of the most advanced technologies and manufacturing in these recent years, allowing users to create complex, custom or patient-specific and analyse the efficiency of the design. Currently, in the market,

Table 1.1 Latest statistics of hip replacement surgeries executed in various countries, categorised by alphabetical order

Country	Registry	Since year	Cumulative THA surgeries*	References
Australia	AOANJRR	1999	Nearly 800,000	[6]
Canada	Canadian Joint Replacement Registry (CJRR)	2016	Nearly 300,000	[7]
Denmark	Danish Hip Arthroplasty Register (DHR)	1995	Over 245,000	[9]
Japan	Japanese Society for Replacement Arthroplasty (JSRA)	2013	Over 282,000	[10–12]
Netherlands	*Landelijke Registratie Orthopedische Implantaten* or Dutch Arthroplasty Register (LROI)	2007	Over 500,000	[13]
New Zealand	New Zealand Joint Registry (NZJR)	1999	Over 348,000 (until 2020)	[14]
Norway	Norwegian Arthroplasty Register (NAR)	1987	Over 260,000	[15]
Scotland	Public Health Scotland (PHS)	2001	Nearly 155,000	[16, 17]
Sweden	Swedish Arthroplasty Registry (SAR)	1979	Over 516,000	[18]
UK	NJR	2003	Over 1,344,000	[4]
USA	AJRR	2012	Over 1,060,000	[5]

*cumulative data of THA surgeries performed (round up to the nearest ten thousand) until 2022, unless stated

there is a lack of three-dimensional (3D) printed hip prosthesis, and the commercial Charnley hip prosthesis has no slot. Aside from that, although there is much research done on finite element analysis (FEA) of hip prosthesis, there is limited research on testing, thus creating a gap in ways to validate the analysis through both FEA and testing. To solve the problems mentioned, this research aimed to evaluate the efficiency of the existing Charnley hip prosthesis by applying AM technology for rapid prototyping and computer-aided design (CAD) modelling and to validate the analysis when doing a comparison of FEA with compressive load testing and hope to push for local production of 3D printed hip prosthesis through this research in future.

1.4   Research Questions

The research questions that this study aims to answer are as follows:

- What method can be used for the development of a hip prosthesis?
- In what ways can the hip prosthesis be developed and analysed?
- What is the testing procedure that can be utilised to evaluate the hip prosthesis?

1.5   Research Objectives

To address the research questions above, the research objectives of this study are as follows:

- To evaluate the existing Charnley hip prosthesis design by identifying the properties and characteristics of existing hip prosthesis using the Pugh chart method.
- To develop a 3D model of a hip prosthesis using CAD modelling and analyse the designed hip prosthesis by the usage of the FEA technique.
- To conduct compressive load testing on the fabricated hip prosthesis by designing and developing a suitable jig and comparing the result of total deformation with FEA.

1.6   Scope and Limitations of Study

This research has a variety of scopes and limitations that it adheres to. Being a biomechanical topic, it is crucial to mention that this research focuses only on the mechanical component of the study. Thus, the analysis conducted may overlook the influence of other factors, such as actual physical strength or the impact of muscle and tissue involved in real-life applications. For this research, the design used was the Charnley hip stem which was later redesigned via CATIA V5R21 software, and the boundary condition parameters used are for standing position. In addition to that, it should be noted that the simulation is conducted under static structural motion via ANSYS 2020 R1 software. The static test is done as one of the primary steps, as it is crucial for evaluation purposes before proceeding to a more complex analysis. Hence, it might affect or limit the analysis when applied to dynamic motion and fatigue analysis scenarios. Furthermore, it is important to highlight that the fabrication of the hip prosthesis is outsourced to our industrial collaborator, who employs the AM 3D printing method. Similarly, the fabrication of the jig is also outsourced. This circumstance thus implies that certain aspects of the manufacturing process and material properties may be beyond the direct control of the study. Apart from this, the fabricated hip prosthesis is only subjected to compressive load testing, which restricts the possibility of comparing other data analysis such as equivalent (von Mises) stress and shear stress within the FEA framework. Also, due to budget constraints associated with fabricating more 3D printed specimens, the compressive load testing was done only twice. These tests were carried out once with the presence of Polymethyl Methacrylate (PMMA) and once without it.

1.7 Significance of Study

The significance of this study revolves around the development of a hip prosthesis that is redesigned and has better analysis, in terms of total deformation, equivalent (von Mises) stress and shear stress, than the existing hip implant, creating an optimum stress distribution and lower stress shielding for the hip implant. Moreover, through collaboration with the industry partner, the design is targeted to be fabricated by the usage of AM technology through rapid prototyping, with the intention of manufacturing it locally. This research is aligned with one of the pillars of IR 4.0 which places emphasis on digital technology, including AM 3D printing technology. Aside from that, it also corresponds with the aim of the Sustainable Development Goals (SDG) 9, which is to build resilient infrastructure, promote inclusive and sustainable industrialisation and foster innovation. In addition to that, the data and information collected in this study may be utilised for further research.

1.8 Layout of Book

This book consists of six significant chapters. This layout provides a summary of each chapter, where each chapter describes a more comprehensive understanding of the problem being studied and helps to gather essential information to develop a hip prosthesis using AM 3D printing technology.

In this chapter, titled "Design and Development of Hip Implant Using Finite Element Analysis (FEA)," introduces the background of the research and explains the reasons behind the interest in conducting it. This chapter outlines the main challenges, research questions, and objectives, as well as the scope and current issues related to hip implants. It ensures that the research stays focused and is completed within the designated timeframe.

In Chap. 2, "Anatomy of Hip and Biomaterial for Hip Implant," the author reviews previous studies related to the topic, covering hip anatomy, total hip arthroplasty (THA), FEA, jig preparation and testing of the fabricated hip prosthesis. These past studies provide a foundation for comparing and analysing the results obtained in the new research.

Chap. 3, "Pugh Chart Method and Finite Element Analysis (FEA) of Hip Implant," details the design and analysis phases involved in developing the new hip implant. It discusses the use of CAD and FEA software, jig preparation and compressive load testing to evaluate the implant and compare the results with simulation data.

Chap. 4 focuses on the "Development of Hip Implant," highlighting the process of creating the hip implant design.

In Chap. 5, "Finite Element Analysis (FEA) of Modified Design of Hip Implant," the results of the FEA on the hip implant and the bone assembly with the implant are presented. This chapter also compares the compressive load testing results of the implant with and without PMMA, discussing the findings in detail with data validation against previous research. Visual aids like graphs and tables are used to help understand the outcomes.

Finally, Chap. 6, "Fundamental of Design and Development of Hip Implant," summarises the entire research process, including data gathering and analysis. It also offers suggestions and recommendations for future research to further strengthen this field.

1.9   Summary

In this chapter briefly introduces the background of the topic, provides the main challenges, research questions and objectives, scopes, limitations and significances of the study and gives the layout of this book. Chapter 2 reviews the topic of the research, which includes hip anatomy, hip replacement surgery, biomaterials in hip arthroplasty on hip prosthesis and bone cement, research gap on this topic.

References

1. About the hip joint, BoneSmart. Accessed 05 March 2020. [Online]. Available: https://bonesm art.org/hip/about-the-hip-joint/
2. M. Merola, S. Affatato, Materials for hip prostheses: a review of wear and loading considerations. Materials **12**(3), 495 (2019). https://doi.org/10.3390/ma12030495
3. J.M. Lee, The current concepts of total hip arthroplasty. Hip Pelvis **28**(4), 191–200 (2016). https://doi.org/10.5371/hp.2016.28.4.191
4. M. Reed, et al., *National Joint Registry (19th Annual Report), National Joint Registry* (2022). Accessed 10 Dec 2022. [Online]. Available: https://reports.njrcentre.org.uk/Portals/0/PDFdow nloads/NJR19thAnnualReport2022.pdf
5. B.D. Springer et al. *American Joint Replacement Registry on Hip and Knee Arthroplasty (9th Annual Report)*. American academy of orthopaedic surgeons (American Joint Replacement Registry) (2022). Accessed 10 Dec 2022. [Online]. Available: https://connect.registryapps.net/ hubfs/PDFsandPPTs/2022AJRRAnnualReport.pdf?hsCtaTracking=e22b6617-7eba-4a95-8113-7aa80eb589d1%7C9509b20f-338c-45c0-aeb2-16a46cffd1d2
6. N. Bergman, et al., *Australian Orthopaedic Association National Joint Replacement Registry— Hip, Knee & Shoulder Arthroplasty (23rd Annual Report)*. Australian Orthopaedic Asso-ciation National Joint Replacement Registry (2022). Accessed 10 Dec 2022. [Online]. Available: https://aoanjrr.sahmri.com/documents/10180/732916/AOA+2022+AR+Digital/f63 ed890-36d0-c4b3-2e0b-7b63e2071b16

7. CIHI, *Hip and Knee Replacements in Canada—CJRR Annual Report 2020–2021*. Canadian Institute for Health Information, 2022. Accessed 10 Dec 2022. [Online]. Available: https://www.cihi.ca/sites/default/files/document/hip-knee-replacements-in-canada-cjrr-annual-report-2020-2021-en.pdf

8. J. Sniderman, A. Khoshbin, J. Wolfstadt, The influence of the COVID-19 pandemic on total hip and knee arthroplasty in Ontario: a population-level analysis. Can. J. Surg. **66**(5), E485–E490 (2023). https://doi.org/10.1503/CJS.016122

9. S. Overgaard, et al., *National årsrapport for 2021 [National Annual Report for 2021]*. Danish Hip Arthroplasty Register (2022). Accessed 11 Dec 2022. [Online]. Available: http://danskhoftealloplastikregister.dk/wp-content/uploads/2022/07/DHR-aarsrapport-2021_Udgivet-2022_offentliggjort-version-1.pdf

10. S. Matsuda, et al., 2013–2017 人工関節登録 報告書 [JSRA 2013–2017 Management Report]. The Japanese Society for Replacement Arthroplasty (2020). Accessed 11 Dec 2022. [Online]. Available: https://jsra.info/data/pdf/report-all-cases.pdf

11. S. Matsuda, et al., 2020 年度症例統計 [JSRA Report 2020], The Japanese Society for Replacement Arthroplasty (2020). Accessed 11 Dec 2022. [Online]. Available: https://jsra.info/data/pdf/report-2020.pdf

12. S. Matsuda, et al., 2021 年度症例統計 [JSRA Report 2021], The Japanese Society for Replacement Arthroplasty (2021). Accessed 11 Dec 2022. [Online]. Available: https://jsra.info/data/pdf/report-2021.pdf

13. I.M.A. de Reus, et al., *LROI Annual Report 2022*, Dutch Arthroplasty Register (2022). Accessed 11 Dec 2022. [Online]. Available: https://www.lroi-report.nl/app/uploads/2022/11/PDF-LROI-annual-report-2022.pdf

14. J. McKie, et al., *The New Zealand Joint Registry 22 Year Report*, New Zealand Joint Registry (2021). Accessed 11 Dec 2022. [Online]. Available: www.nzoa.org.nz/nzoa-joint-registry

15. O. Furnes, et al., *Norwegian National Advisory Unit on Arthroplasty and Hip Fractures*. Norwegian Arthroplasty Register (2022). Accessed 11 Dec 2022. [Online]. Available: https://helse-bergen.no/seksjon/Nasjonal_kompetansetjeneste_leddproteser_hoftebrudd/SharepointDocuments/Rapport/Report2022english.pdf

16. M. Moran, et al., *Scottish Arthroplasty Project Report 2022* (2022). Accessed 10 Dec 2022. [Online]. Available: www.publichealthscotland.scot

17. M. Moran, et al., *Scottish Arthroplasty Project Report 2022—Main Report*. Scottish Arthroplasty Project (SAP) (2022). Accessed 11 Dec 2022. [Online]. Available: https://publichealthscotland.scot/publications/scottish-arthroplasty-project/scottish-arthroplasty-project-13-september-2022/dashboard/

18. A. W-Dahl, et al., *SAR Annual Report 2022*. Swedish Arthroplasty Register (2022). Accessed 11 Dec 2022. [Online]. Available: https://registercentrum.blob.core.windows.net/sar/r/SAR-Annual-Report-2022_EN-HkgQE89Nus.pdf

19. H. Kim et al., Variations in hip fracture inpatient care in Japan, Korea, and Taiwan: an analysis of health administrative data. BMC Health Serv. Res. **21**, 694 (2021). https://doi.org/10.1186/s12913-021-06621-y

20. W.L. Healy, R. Iorio, A.J. Clair, V.D. Pellegrini, C.J. Della Valle, K.R. Berend, Complications of total hip arthroplasty: standardized list, definitions, and stratification developed by the hip society. Clin. Orthop. Relat. Res. **474**(2), 357–364 (2016). https://doi.org/10.1007/s11999-015-4341-7

21. M. A. H. Abdullah, et al., Annual report of national orthopaedic registry Malaysia (NORM) hip fracture 2009. *National Orthopaedic Registry of Malaysia* (2010), pp. 1–35. Accessed 06 Oct 2023. [Online]. Available: https://www.crc.gov.my/wp-content/uploads/documents/hip_norm.pdf

22. T. Ong et al., The current and future challenges of hip fracture management in Malaysia. Malays. Orthop. J. **14**(3), 16–21 (2020). https://doi.org/10.5704/MOJ.2011.004

23. G.D. Mandavgade, T.R. Deshmukh, Hip implant: CAD modelling and static analysis. Trends. Biomater. Artif. Organs. **33**(4), 1–7 (2019) Accessed 20 April 2020. [Online]. Available: https://sipnaengg.ac.in/wp-content/uploads/2020/09/GDM-1_compressed.pdf

Chapter 2
Anatomy of Hip and Biomaterial for Hip Implant

2.1 Introduction

This chapter reviews past research to gather information on how the research is conducted. It also includes discussions in detail on hip anatomy, hip replacement surgery, biomaterials in hip prosthesis and bone cement, and research gap on this topic.

2.2 Hip Anatomy

The human body has several joints and one of the largest amongst them is the hip. The hip anatomy includes the pelvis, acetabulum, acetabulum labrum (cartilage), femoral head, femoral neck and thighbone (femur) as can be seen in Fig. 2.1. The ball and socket is also known as the femoral head and acetabulum, respectively [1]. The ligaments provide stability as they hold the joint together [2]. A healthy joint refers to the femoral head that fits into the acetabulum of the hip bone which allows the movement of the leg [1] in a large range of motion in doing daily activities, for instance walking, running, climbing and seating [2]. Figure 2.2 illustrates the types of motion of the hip joint: flexion, extension, abduction, adduction, lateral (external) and medial (internal) rotations [3].

2.3 Hip Replacement Surgery

Today, one of the most successful orthopaedic surgeries done over the past 30 years is THA, also referred to as THR. THA acts as a pain reliever, restores function and enables humans' quality of life to be improved. The surgery helps patients who

© The Author(s), under exclusive license to Springer Nature Singapore Pte Ltd. 2025   9
N. S. Hamizan et al., *Hip Prosthesis*,
SpringerBriefs in Applied Sciences and Technology,
https://doi.org/10.1007/978-981-96-1470-7_2

Fig. 2.1 Hip anatomy [4][1]

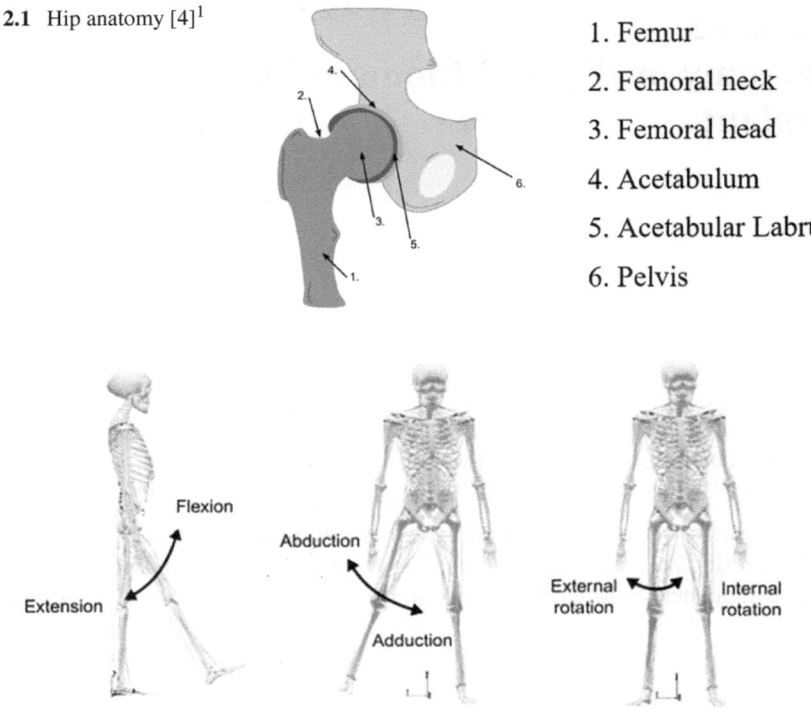

1. Femur

2. Femoral neck

3. Femoral head

4. Acetabulum

5. Acetabular Labrum

6. Pelvis

Fig. 2.2 Hip joint movement [3][2]

suffer damage, bone fractures or dysfunctional hip joints, usually due to degenerative diseases such as osteoarthritis [1]. The procedure is done as the removal of the damaged part and replacement of an artificial implant, functioning as a normal hip joint. Different kinds of procedures of hip arthroplasty surgery include traditional, minimally invasive, partial and revision surgeries. Minimally invasive were performed by using one or two smaller incisions [3].

In a standard hip replacement procedure, the prosthesis consists of several components, each serving a specific role [5]. The hip prosthesis includes several parts which are the femoral stem, insert, femoral heads and acetabular components as in Fig. 2.3. The materials of these pieces could be produced in the form of metal, ceramic, plastic or combinations of those (hybrid). Firstly, the stem is inserted into the femur or thigh bone, providing stability and support. Secondly, the acetabular cup is inserted into the pelvic bone, creating a socket for the ball of the joint. The femoral head fits onto the end of the stem, forming the articulating surface of the joint. Finally, the insert

[2] Reprinted from Computational Modelling of Biomechanics and Biotribology in the Musculoskeletal System, 2, X. Zhang, Computational modelling of biomechanics for an artificial hip joint, 517–546, 2021, with permission from Elsevier.

Fig. 2.3 Illustration of a hip prosthesis embedded in femur in THA surgery [6][3]

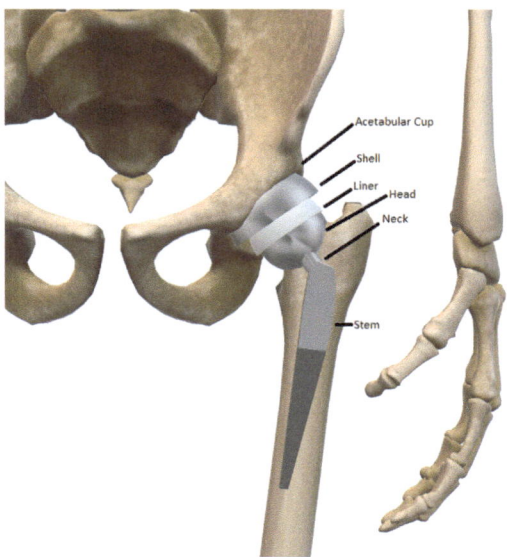

or liner, which acts as the new cartilage, is inserted into the cup, allowing for smooth movement and reducing friction within the joint. Together, these components work harmoniously to improve the overall function of the hip joint and alleviate discomfort for individuals who have undergone hip replacement surgery.

Every hip implant system is characterised by its unique device design features, including factors such as size, shape, material and dimensions. The selection of a specific hip prosthesis for a patient is a decision that is made by the orthopaedic surgeon based on individual factors, taking into consideration the maximum benefit and minimising any potential complications for each patient.

Hip replacement surgery involves the usage of artificial components to replace the entire hip structure. During the procedure, surgeons insert a stem into the patient's femur for stability, replace the head of the femur with a ball and replace the natural socket in the hip joint with an artificial cup, as shown in Figs. 2.4 and 2.5.

Currently, there are four types of THR devices available with different bearing surfaces [9]. The first type is metal-on-polyethylene (PE), whereby the ball component is made of metal and the socket is made of plastic (PE) or has a plastic lining. The second type is ceramic-on-PE, whereby the ball is made of ceramic and the socket is made of plastic or has a plastic lining. The third type is ceramic-on-ceramic, whereby both the ball and the socket have a ceramic lining. Lastly, there is ceramic-on-metal, whereby the ball is made of ceramic and the socket has a metal lining. These different combinations of materials provide surgeons with options when selecting the most suitable hip replacement device for a patient's specific needs.

Fig. 2.4 Preparation of hip
prosthesis insertion (*Source
AO surgery reference, www.
aosurgery.org*) [7][4]

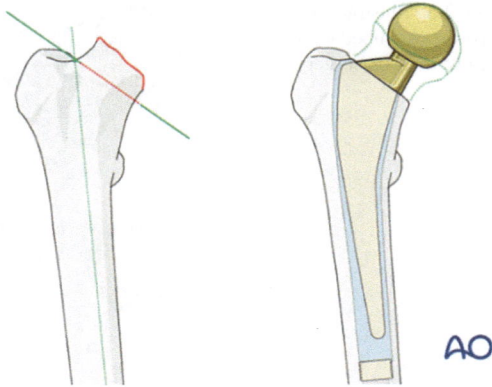

Fig. 2.5 Example of an
X-ray of an implanted hip
prosthesis within the femur
[8][5]

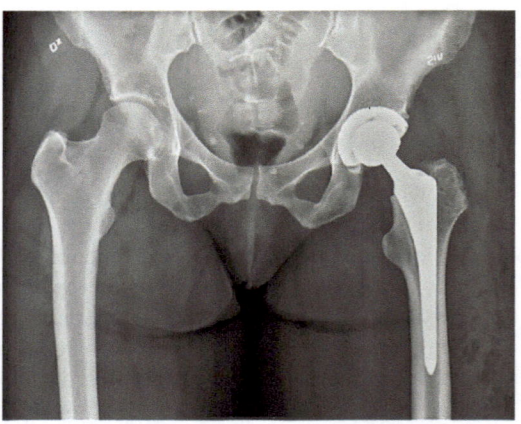

The three common techniques available for THA are cemented, uncemented and
hybrid (a combination of cemented and uncemented implants) [3]. Special bone
cement is commonly used to hold hip prosthesis in place, but some surgeons use a
cementless fixation technique. Implants designed without the need for cement feature
a textured surface that promotes bone integration which allows the bone to grow onto
the implant and secure it. A hybrid THR involves implanting the cup without cement
and setting the ball in place with cement.

A typical, uncomplicated THR surgery involves several steps [10]. First, the
patient is prepared in the operating room by receiving an IV line and possibly a
urinary catheter. Anaesthesia is then administered, either through general or regional
anaesthesia. The surgeon makes an incision in an appropriate location to access the

---

[4] Copyright by AO Foundation, Switzerland.

[5] "X-ray of pelvis with total arthroplasty" by Mikael Häggström, used under public domain CC0
1.0 Universal.

hip joint depending on the chosen approach. The head of the femur is removed, and an opening is made at the top to insert the ball prosthesis into the femur. The damaged cartilage in the socket part of the joint is removed, and a new cup component is attached. The surgeon tests the functionality of the joint by moving the patient's leg. Tissue layers are closed, and the outermost layer of skin is sealed with surgical glue. After a few hours in the recovery room, the patient is transferred to a hospital room. The duration of the surgery is typically around two hours, but it may vary depending on the type of procedure and any complications. Imaging, such as X-rays, may be conducted after the surgery and during recovery to ensure proper healing.

2.3.1 Existing Hip Prosthesis

The market for hip prosthesis encompasses several leading companies that operate and offer a range of options to patients in need of hip replacement surgery [11]. Some notable companies in this field include Zimmer Biomet [12], DePuy Synthes Companies (Johnson & Johnson Services, Inc.) [13], Smith & Nephew [14], Stryker Corporation [15], Exactech, Inc. [16], Gruppo Bioimpianti s.r.l, [17], Conformis [18], DJO Global [19], Microport Scientific Corporation [20] and B. Braun Medical [21].

2.3.2 Bone Cement

PMMA, typically called as bone cement, is a versatile material that is used widely for implant fixation in artificial joint procedures. It serves as a space filler, enabling the creation of a secure and tight space in attaching the bone and the prosthesis. The utilisation of bone cement in joint replacement surgeries offers several notable advantages. These advantages [22] include addressing bone porosity as the bone cement has the ability to address bone conditions like osteoporosis, thus enhancing the stability of the implant and reducing the risk of implant loosening, decreased post-surgical infection risk whereby the addition of a small amount of antibiotic material to the bone cement can help to inhibit the growth of bacteria, and finally rapid hardening time for secure fixation as the bone cement dries within 10 min of application, allowing the surgeon and patient to have the confidence that the prosthetic joint components are firmly in place.

The use of bone cement in joint replacement surgeries, although beneficial, does have some potential drawbacks and risks that should be considered [22]. These include the loosening of the artificial joint over time as the bone cement may degrade, which may prompt the need for another joint replacement surgery (revision surgery), irritation and inflammation to the surrounding soft tissue due to the bone cement debris that breaks off, resulting in discomfort and pain around the joint area. Also, risk of cement embolism, whereby the cement can enter the bloodstream and subsequently reach the lungs, a condition that can be life-threatening, in rare cases, uncertainty on the exact frequency of these complications occurs following specific types of joint replacement surgeries. Bone cement debris does not always cause symptoms in patients. Bits of cement debris can be removed arthroscopically to alleviate or prevent symptoms, and allergic reactions to the bone cement for some patients and must undergo a second surgery to remove the cement and prosthesis to relieve discomfort and adverse reactions caused by the cement, in rare cases.

2.3.3 Hip Replacement in Malaysia

The hip joint, being a remarkable ball and socket joint, plays a crucial role in supporting the body's weight and facilitating smooth movement. However, due to various factors such as disease or the natural effects of wear and tear, the hip joint can become damaged, leading to discomfort and limited mobility. It is important to note that the decision to undergo a hip replacement is made on a case-by-case basis, taking into consideration factors such as the extent of joint damage, the patient's overall health and their specific needs and goals. A thorough evaluation by a qualified orthopaedic specialist is crucial in determining the most appropriate course of action for everyone. There are four types of hip replacement surgery that can be done in Malaysia which are THR, partial, minimally invasive and revision hip replacement surgeries.

In cases where only a portion of the joint is affected, a partial hip replacement surgery can be considered as a viable solution. During a partial hip replacement surgery, the orthopaedic surgeon aims to address the damaged portion of the joint while preserving the integrity of the hip socket. Unlike a THR, where both the head of the femur bone and the hip socket are replaced, in a partial hip replacement, the hip socket is usually left intact [23, 24]. In some cases, an alternative approach may be employed to avoid cutting the top of the femur bone. This involves fitting a specialised device over the bone, which can provide support and stability without the need for removing the top of the femur. By using this method, the surgeon can achieve the desired outcome while minimising the extent of the surgical procedure.

Minimally invasive hip replacement is an innovative surgical technique that offers several potential benefits as compared to traditional hip replacement surgery. With this approach, a skilled surgeon can perform the hip replacement procedure through one or two small incisions, as opposed to the larger incisions required in the conventional method [25, 26]. By utilising smaller incisions, minimally invasive hip replacement aims to minimise trauma to the surrounding tissues, muscles and ligaments. This approach may result in reduced post-operative pain and discomfort for patients. Another advantage of minimally invasive hip replacement is the potential for faster recovery and a quicker healing process [25, 26]. However, it is important to note that despite the potential benefits, minimally invasive hip replacement is a relatively new technique, and its long-term outcomes are still being evaluated. As such, there

Table 2.1 Cost of hip replacement in Malaysia [24]

Procedure	Price in USD (MYR conversion)*	
	Minimum	Maximum
THR	$ 4,600 (RM 21.7 k)	$ 6,500 (RM 30.7 k)
Minimally invasive	$ 7,700 (RM 36.3 k)	$ 8,500 (RM 40.1 k)
Partial	$ 5,100 (RM 24.1 k)	$ 7,100 (RM 33.5 k)
Revision	$ 10,000 (RM 47.2 k)	$ 15,000 (RM 70.8 k)

*The conversion rate of USD 1 to RM 4.72 was provided as a reference for the price listed in the table that might involve currency exchange

is currently a lack of reliable studies directly comparing the outcomes of minimally invasive hip replacement to the traditional method.

Revision hip replacement surgeries are performed to replace an artificial implant used in hip replacements that wears down or fails over time which leads to deterioration, with a new one [27]. It is important to recognise that revision hip replacements are typically more complex and challenging as compared to the initial hip replacement surgery. This complexity arises from several factors, such as the presence of scar tissue, changes in bone structure and potential bone loss resulting from the previous surgery. While revision hip replacements aim to address issues with the existing implant, it is worth noting that the outcomes of revision surgeries may not be as favourable as those of the primary hip replacement [28, 29]. Another important consideration is the increased risk of complications with each subsequent revision surgery, thus can contribute to a higher risk of complications, including infection, implant instability, bone fractures and limited range of motion.

In Malaysia, the average cost of different types of hip replacement surgeries available is tabulated in Table 2.1 as follows:

2.4 Biomaterials in Hip Arthroplasty

The development of biomaterials has had a significant impact on the medical field over the decades. A biomaterial is a substance that is designed with the purpose of interacting with the body. Biomaterials are used in medical applications to support, enhance or replace damaged tissue or a biological function. They can be natural or synthetic. Metals, ceramics, plastics, glass and even living cells and tissues can be used to create biomaterials. One of the medical field applications that utilise biomaterials is medical implants, including heart valves, stents and grafts; artificial joints, ligaments and tendons; hearing loss implants; dental implants; and nerve stimulation devices.

2.4.1 Hip Prosthesis

Hip replacement parts weigh roughly about 3 to 5 pounds (1.3 to 2.3 kg). Patients may gain a few pounds of body weight due to hip replacement surgery as the bone removed during surgery weighs a little less [30]. For the present design, by using the AM 3D printing technology method, it is expected for the hip implant to be lighter than the existing design. Hence, a lighter weight is preferable for the invention. The weight of a hip prosthesis differs by the type of materials used, design or fabrication method. That being the case, it is hard to obtain specific ranges of weights for various materials of hip implants in the market.

Currently, patients are receiving their hip prosthesis from overseas markets such as the USA, the UK, Europe and China. The duration to import the hip prosthesis took a minimum of at least a month through the shipping process. The invention thus will help in reducing the time and cost as the hip prosthesis will be fabricated locally. This will especially be helpful in trauma cases patient who requires immediate replacement and treatment as rapid prototyping of this invention may only take hours or days to be produced instead of months of waiting for the ready-made hip prosthesis to be imported.

Mechanical properties such as hardness, tensile strength, Young's modulus and elongation are used to determine the type of metallic material. An implant fracture due to a mechanical failure is related to biomechanical incompatibility [31]. For this reason, it is expected that the material employed to replace the bone has similar mechanical properties to that of a bone. The Young's modulus of a bone varies in a range of 4 to 30 GPa depending on the type of the bone and the direction of measurement [31]. At times, the metallic materials used for biomedical applications are titanium-based alloys (Ti–6Al–4 V), cobalt-chromium alloys (CoCrMo), stainless steel and others [31].

Weighs about half as much as steel but is 30% stronger, titanium is a metal known for its strength and lightness [32, 33]. Owing to their outstanding properties, great tensile strength, flexibility and high corrosion resistance, titanium and titanium alloys are some of the most used implant materials for biomedical applications. Due to its resistance to corrosion from bodily fluids, it is regarded as the most biocompatible metal as it is not harmful or toxic to living tissue [34]. It exhibits a unique combination of strength and biocompatibility, which enables its use in medical applications and accounts for its extensive use as an implant material in the last 50 years. Aside from that, it has less lack of fusion defects. By defining porosity within the implant, it can also improve the stress shielding issues [34].

This metal's ability to physically bond with bone gives titanium an advantage over other materials that require the use of an adhesive to remain attached. Titanium implants last longer, and much larger forces are required to break the bonds that join them to the body compared to their alternatives [34]. Titanium alloys are commonly used in load-bearing implants and are significantly less stiff than stainless steel or cobalt-based alloys [34]. The medical industry uses titanium for the human body because it presents a similar density to bone and due to its high strength-to-low weight ratio [33].

In engineering applications, steels continue to be the most common metallic material to be used. Stainless steel 316L is an example of the application of these alloys as a biomaterial. It can be used in the manufacture of elements such as femoral stems and heads, combining good mechanical strength and corrosion resistance [31]. However, the main advantage of stainless steel in relation to other metallic materials is the cost–benefit ratio. In general, stainless steels are versatile in terms of properties [31]. There is a linear relationship between the manufacturing process, structure and properties. For example, forging is a bulk deformation process in metal working commonly employed in the manufacture of stainless steel prosthesis [31]. In a material, the formation of structures, or grain sizes, increases its mechanical strength and is dependent upon the compressive loads applied to it. However, when it is intended to improve properties for application as biomaterial, other metallic materials stand out, amongst them titanium alloys and cobalt alloys [31]. These nonferrous metal materials, despite having higher processing costs than stainless steel, provide these materials with gains in mechanical strength and corrosion resistance, providing a longer useful life for prosthesis made with such materials [31].

Cobalt (Co)-based implants have higher wear resistance as compared to Ti alloys, which warrants their extensive use in artificial hip joints, where the direct contact between the femoral head and the bone or plate over time may lead to wear [35]. Clinically, Co–Cr–Mo is one of the most used alloys due to a favourable combination of high strength and high ductility. Wrought Co–Cr alloys containing nickel (Ni), e.g., Co–Ni–Cr–Mo, offer greater strength than cast Co–Cr alloys. However, it is only used in those applications where extra strength is needed since Ni is potentially toxic [35]. The elastic modulus of Co–Cr alloys is also higher than that of commercially pure Ti or Ti alloys. The yield strength and tensile strength of Co–Cr alloys are in the range of 448 MPa to 1,606 MPa and 655 MPa to 1,896 MPa, respectively, whereas Ti–6Al–4 V has a yield strength of 896 MPa to 1,034 MPa and the tensile strength of 965 MPa to 1,103 MPa. Compared to that of bone, the Co–Cr alloys have higher elastic modulus and greater density and stiffness, which leads to greater stress shielding than in the case of Ti and Ti alloys or Mg [35]. The biocompatibility of and osseointegration capacity of Co–Cr is also lower than that of Ti. Therefore, it is typical for Ti to be used for elements that will be in direct contact with the bone, e.g., screws, and Co–Cr to be the preferred material for those elements that do not interface with the bone, e.g., rods in spinal fixation in clinical settings [35]. However, these structures result in metal corrosion, specifically considerable mechanically assisted crevice corrosion, and shredding at the site of the contact between Co–Cr and Ti, which is subject to a significant frictional load [35]. The tissue indeed is in the proximity of the interface between these materials in THA, knee implants and spinal fixation most commonly experience metallosis, which is a medical condition disorder caused by deposition and build-up of metal debris in the soft tissues of the body of patients. Due to the superior wear resistance of Co–Cr alloy, Ti is a major source of metallic debris in such constructs [35].

2.4.2   Bone Cement

Bone cement can be defined as biomaterials obtained by mixing a powder phase and a liquid phase, which can be moulded and implanted as a paste and can be set once implanted within the body. They are widely utilised in several orthopaedic surgery applications. At present, the most frequently used bone cement can be classified into two groups, namelyacrylic bone cement (ABC), calcium phosphate bone cement (CPC) [36] and glass polyalkenoate cement (GPC). ABC is polymer whereas CPC is ceramic. CPCs are intrinsically porous materials, so their strength is in general lower than that of ABC. Also, due to its nature as ceramic, which is brittle, CPC is not suitable as a fixation for the articular prosthesis. GPC is formed by an acid–base reaction between a water-soluble polyacrylic acid (PAA) and an acid-degradable fluoro-alumino-silicate bioactive glass [37]. Commonly referred to as glass ionomer cement, GPC is mainly used in dentistry, and ear, nose and throat (ENT) applications [38, 39].

ABC is based on PMMA which is accepted as a biocompatible polymer when cured. The bone cement is prepared by mixing two components, a liquid phase and a powder phase [36]. For the cement to be produced, it is necessary for the liquid monomer to wet the powder particles of PMMA. The hardening process of bone cement can be affected by several factors. The polymerisation process is strongly dependent upon the environmental temperature at which the reaction takes place [36]. In order to be able to compare results, the American Society for Testing and Materials (ASTM) and International Organization for Standardization (ISO) standards on ABC state that the dough time and the setting time evaluation should be conducted at a room temperature of $23 \pm 1\,^{\circ}\text{C}$ and relative humidity of $50\% \pm 10\%$ [36]. Mixing technique also plays a role in affecting the process. Vacuum mixing hardens faster than manual mixing [40]. One flow of rotation of mixing, either clockwise or anticlockwise, is preferable to avoid any air bubbles from being trapped inside the cement mixture.

The bone cement in the market may or may not contain gentamicin (antibiotics). These antibiotics are used to help in preventing surgical-site infection [40]. Standardised cement specimens are made from 40 g PMMA loaded with 1 g antibiotics [41]. Hence, it is recommended for patients to undergo an allergy test prior to the surgery, albeit it is not very common to have allergies to components of bone cement, to avoid any implications [40].

2.5   Research Gap

While 3D printing technology has been explored and researched extensively in the field of orthopaedic implants, including hip prosthesis, the widespread availability of 3D printed hip prosthesis in the commercial market has been relatively limited. However, there are several companies and research institutions that have explored the use of 3D printing to manufacture hip prosthesis. The first 3D printed hip prosthesis

Fig. 2.6 Example of a
3D-printed acetabular cup
prosthesis [47][6]

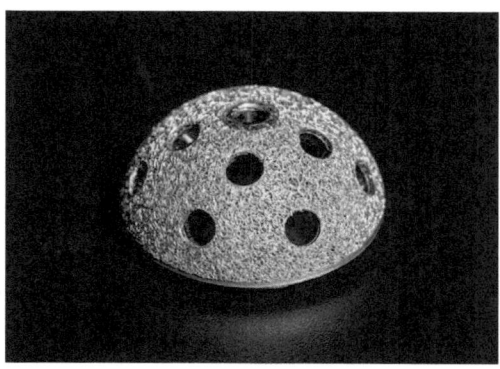

was the Delta-TT Cup, which stands for trabecular titanium acetabular cup, created by Dr. Guido Grappiolo back in 2007, and until 2018, it has been implanted in more than 100,000 patients [42–45]. Another case involving a 3D printed hip prosthesis was for custom titanium acetabular prosthetics due to a huge defect in the patient's hip with no proximal femur and no pelvis that made it impossible to use existing implants [46]. Figure 2.6 shows an example of a 3D printed acetabular cup prosthesis.

While the 3D printed acetabular cup has been progressing rapidly and implanted in patients, the same cannot be said with the femoral stem. Currently, there is no known case that uses any available 3D printed femoral stem hip prosthesis implanted in patient that was documented, thus creating a gap to explore more on the 3D printed femoral hip stem.

Apart from that, although numerous studies have investigated the FEA performance of hip implants, there remains a notable gap in analyses through both FEA and testing methodologies, thus amplifying the research void. Additionally, it aims to bridge the gap between FEA and testing by conducting comparative analyses with compressive load testing.

Table 2.2 provides some previous research highlights related to this research. It is a bit challenging to find FEA and testing on hip prosthesis as there are some that use various designs, or analyse only on the hip implant, or conduct tests using 3D printed hip prosthesis. However, Čolić et al. [48, 49] and Suyitno et al. [50] have done research on both FEA and testing that are pretty like this research. Čolić et al. [48, 49] presented in their two research papers that the static test had a 15% difference to the FEA result, whereas Suyitno et al. [50] obtained 1.5 mm deformation at 2.0 kN force load but with testing higher than the FEA results. Also, Arabnejad et al. [51] came out with porous structures that can reduce stress shielding more than a fully solid hip prosthesis.

---

[6] "Pelvic Component for Hip Joint Replacement (5185) (18,494,294,741)" by The U.S. Food and Drug Administration, used under public domain CC0 1.0 Universal.

Table 2.2 Previous research highlights

Year	Scientist(s)	Material/Part	FEA	Testing	3D print	Findings
2021	Corona-Castuera et al. [52]	Stainless steel stem	Bone assembly	N/A	DMLS	Triply periodic minimal surfaces (TPMS)
2020	Ahmad et al. [53]	Stainless steel stem, PMMA	Different cement mantle stair climb and walk	N/A	N/A	Fully cemented provides minimum displacement and minimises the risk of implant loosening
2019	Tabaković et al. [54]	Stem & ABS filament cup	Custom design and bone assembly	CAD inspection	N/A	Large deviations in FDM and CNC machining give better result
2017	Suyitno and Sutomo [50]	Ti grade 2 stem, SS316L ball socket, UHMWPE cup & ball, PMMA resin	Bone assembly normal walk	Static test Normal walk $F = 2.81$ kN Rate 2.8 mm/min With PMMA	FDM	Total deformation of 2.0 kN at 1.5 mm when compared to FEA and simulation, the testing also able to withstand up to 2.8 kN
2017	Čolić et al. [49]	Ti–6Al–4 V stem	Implant only Normal walk	Static test Normal walk, DIC $F = 2.3$ kN	N/A	Correspond until 6 kN Comparison about 15% difference where FEA is larger
2016	Arabnejad et al. [51]	Ti alloy stem	Bone assembly	DIC On bone and bone assembly	SLM	Bone loss from 27% in fully solid to 8% in fully porous, reduces stress shielding

(continued)

Table 2.2 (continued)

Year	Scientist(s)	Material/Part	FEA	Testing	3D print	Findings
2016	Čolić et al. [48]	Ti–6Al–4 V stem	Implant only Slow walk and trip	N/A	N/A	Maximum equivalent (von Mises) stress is between 256 MPa on slow walk, and 535.5 MPa when involving trip
2011	Paciago et al. [55]	N/A	Existing & custom 2 stems	N/A	N/A	Highest stress is at the neck maximum equivalent (von Mises) stress = 427 MPa & 450 MPa Maximum displacement = 3.32 & 1.8 mm (50% reduction) Minimum factor of safety (FOS) = 3.94 & 4.33
2010	Jun and Choi [56]	N/A	Custom, patient-specific	N/A	N/A	Exclusive and practical but needed a detailed validation process before clinical use

2.6 Summary

This chapter reviews hip anatomy, hip replacement surgery, biomaterials in hip arthroplasty on the hip prosthesis and bone cement and research gap on this topic. Chapter 3 will summarise Pugh chart method, FEA, preparation of jig and compressive load testing of the hip prosthesis.

References

1. T. Terry, B. Kim, *Hip Replacement*, ConsumerNotice. Accessed 29 Oct 2020. [Online]. Available: https://www.consumernotice.org/drugs-and-devices/hip-replacement/
2. What Does a Hip Do?, *BrainLab*. Accessed 29 Oct 2020. [Online]. Available: https://www.brainlab.org/get-educated/hip/hip-anatomy/what-does-a-hip-do/
3. X. Zhang, Computational modelling of biomechanics for an artificial hip joint. In: *Computational Modelling of Biomechanics and Biotribology in the Musculoskeletal System*, 2nd ed., Z. Jin, J. Li, Z. Chen, (eds.) (Woodhead Publishing, 2020), ch. 20, pp. 517–546. https://doi.org/10.1016/B978-0-12-819531-4.00020-1
4. Kcotton15, *File:Ball and Socket Joint (Hip joint).svg*. Wikimedia Commons. Accessed 24 Sept 2024. [Online]. Available: https://commons.wikimedia.org/wiki/File:Ball_and_Socket_Joint_(Hip_joint).svg
5. What are hip and knee replacement implants made of? American association of hip and knee surgeons. Accessed 27 Jan 2022. [Online]. Available: https://hipknee.aahks.org/wp-content/uploads/2019/01/what-are-hip-and-knee-replacement-implants-made-of-AAHKS.pdf
6. M. Shafi, File:Hip Prostesis.png. *Wikimedia Commons*. Accessed 24 Sept 2024. [Online]. Available: https://commons.wikimedia.org/wiki/File:Hip_Prostesis.png
7. T. Apivatthakakul, J.K. Oh, Total hip arthroplasty for Femoral neck and head fractures with hip dislocation. *AO Surgery Reference*. Accessed 27 Jan 2022. [Online]. Available: https://surgeryreference.aofoundation.org/orthopedic-trauma/adult-trauma/proximal-femur/femoral-neck-and-head-fracture-with-hip-dislocation/total-hip-arthroplasty
8. M. Häggström, File:X-ray of pelvis with total arthroplasty.jpg. *Wikimedia Commons*. Accessed 24 Sept, 2024. [Online]. Available: https://commons.wikimedia.org/wiki/File:X-ray_of_pelvis_with_total_arthroplasty.jpg
9. L. Venkataraman, S. Shroff, Hip replacement/hip arthroplasty. *Medindia*. Accessed 27 Jan 2022. [Online]. Available: https://www.medindia.net/surgicalprocedures/hip-replacement.htm
10. Hip Replacement Surgery, *Johns Hopkins Medicine*. Accessed 27 Jan 2022. [Online]. Available: https://www.hopkinsmedicine.org/health/treatment-tests-and-therapies/hip-replacement-surgery
11. FortuneBusinessInsights, *Hip Replacement Market Size, Share & Industry Trends (2021–2028)*. (Fortune Business Insights, 2021). Accessed 03 Aug 2022. [Online]. Available: https://www.fortunebusinessinsights.com/industry-reports/hip-replacement-implants-market-100247
12. ZimmerBiomet, Hip replacement products. *Zimmer Biomet*. Accessed 03 Aug 2022. [Online]. Available: https://www.zimmerbiomet.com/en/products-and-solutions/specialties/hip.html#four
13. DePuySynthes, *Hip Replacement and Reconstruction*. (Johnson & Johnson MedTech). Accessed 03 March 2023. [Online]. Available: https://www.jnjmedtech.com/en-EMEA/specialty/hip
14. Smith&Nephew, *Hip Arthroplasty*. Smith & Nephew. Accessed 03 March 2023. [Online]. Available: https://www.smith-nephew.com/en/health-care-professionals/products
15. Stryker, *Hip Implants*. Stryker. Accessed 03 March 2023. [Online]. Available: https://www.stryker.com/us/en/portfolios/orthopaedics/joint-replacement/hip.html
16. Exactech, *Hip Replacement Products*. Exactech. Accessed 03 March 2023. [Online]. Available: https://www.exac.com/hip/
17. GruppoBioimpiantiGlobal, *Hip Replacement*. Gruppo Bioimpianti Global. Accessed 03 March 2023. [Online]. Available: https://bioimpianti.it/en/patients/hip-prosthesis/
18. Conformis, *Hip Replacement Systems*. Conformis. Accessed 03 Aug 2022. [Online]. Available: https://www.conformis.com/surgeon-resource-center/products/hips/
19. DJOGlobal, *Hip Solutions*. DJO Global. Accessed 03 March 2023. [Online]. Available: https://enovis.com/surgical/hip
20. MicroPortScientificCorporation, *HIPS*, MicroPort Scientific Corporation. Accessed 03 March 2023. [Online]. Available: https://microport.com/healthcare-professional/orthopedics/hips

21. B. Braun, *Hip Arthroplasty*. B. Braun SE. Accessed 03 Aug 2022. [Online]. Available: https://www.bbraun.com/en/products-and-solutions/therapies/orthopaedic-surgery/hip-arthroplasty.html
22. R. Vaishya, M. Chauhan, A. Vaish, Bone cement. J. Clin. Orthop. Trauma. **4**(4), 163 (2013). https://doi.org/10.1016/J.JCOT.2013.11.005
23. E. Smiley, R. Zimlich, E. Stem, S. Burden, *Partial Hip Replacement*. Lowcountry orthopaedics. Accessed 06 Oct 2023. [Online]. Available: https://www.lowcountryortho.com/hip-knee/partial-hip-replacement/
24. *Hip Replacement in Malaysia*. Health-tourism. Accessed 27 July 2022. [Online]. Available: https://www.health-tourism.com/hip-replacement-surgery/malaysia/
25. S. Sporer, *Minimally Invasive Hip Replacement Versus Traditional Hip Replacement*. Arthritis-health. Accessed 06 Oct 2023. [Online]. Available: https://www.arthritis-health.com/surgery/hip-surgery/minimally-invasive-hip-replacement-vs-traditional-hip-replacement
26. J.R.H. Foran, T.W. Throckmorton, *Minimally Invasive Total Hip Replacement*. OrthoInfo. Accessed 06 Oct 2023. [Online]. Available: https://orthoinfo.aaos.org/en/treatment/minimally-invasive-total-hip-replacement/
27. J.R.H. Foran, S.J. Fischer, *Total Hip Replacement*, OrthoInfo from the American Academy of Orthopaedic Surgeons. Accessed 29 Oct 2020. [Online]. Available: https://orthoinfo.aaos.org/en/treatment/total-hip-replacement/
28. VersusArthritis, *Hip Replacement Surgery* (2020). Accessed 27 Jan 2021. [Online]. Available: https://versusarthritis.org/media/23141/hip-replacement-surgery-information-booklet.pdf
29. YorkOrthopaedics, *Revision Hip Replacement*. York Orthopaedics. Accessed 06 Oct 2023. [Online]. Available: https://yorkorthopaedics.co.uk/hip/revision-hip-replacements/
30. *Parts and Materials for Hip Replacement*. Hip and Knee. Accessed 27 Jan 2022. [Online]. Available: https://hipandknee.com/hip-surgery/about-hip-replacement/parts-materials/
31. G.A. dos Santos, The importance of metallic materials as biomaterials. Adv Tissue Eng Regenerative Med: Open Access **3**(1), 300–302 (2017). https://doi.org/10.15406/atroa.2017.03.00054
32. *All About Titanium Aerospace Metal*, ASM Aerospace Specification Metals, Inc. Accessed 27 Jan 2022. [Online]. Available: https://www.aerospacemetals.com/all-about-titanium.html
33. *Top 7 Uses of Titanium—Titek*. Titanium Specialists. Accessed 27 Jan 2022. [Online]. Available: https://titek.co.uk/top-7-uses-of-titanium/
34. L. Zhang, Titanium is the perfect metal to make replacement human body parts. *The Conversation*. Accessed 29 Jan 2022. [Online]. Available: https://theconversation.com/titanium-is-the-perfect-metal-to-make-replacement-human-body-parts-115361
35. K. Prasad et al., Metallic biomaterials: current challenges and opportunities. Materials **10**(8), 884 (2017). https://doi.org/10.3390/ma10080884
36. M.-P. Ginebra, E.B. Montufar, Cements as bone repair materials. In: *Bone Repair Biomaterials*, 2nd ed., K.M. Pawelec, J.A. Planell, (Eds.), (Woodhead Publishing, 2019), ch. 9, pp. 233–271. https://doi.org/10.1016/B978-0-08-102451-5.00009-3
37. P. Niranjan, et al., Injectable glass polyalkenoate cements: evaluation of their rheological and mechanical properties with and without the incorporation of lidocaine hydrochloride voltammetric determination of lidocaine and its toxic metabolite in pharmaceutical formulation and milk using carbon paste electrode modified with C18 silica injectable glass polyalkenoate cements: evaluation of their rheological and mechanical properties with and without the incorporation of lidocaine hydrochloride. Biomed. Phys. Eng. Express. **4**, 27002 (2018). https://doi.org/10.1088/2057-1976/aa952b
38. B.A. Khader, D.J. Curran, S. Peel, M.R. Towler, Glass polyalkenoate cements designed for cranioplasty applications: an evaluation of their physical and mechanical properties. J. Funct. Biomater. **7**(2), 8 (2016). https://doi.org/10.3390/JFB7020008
39. A.M.F. Alhalawani, D.J. Curran, D. Boyd, M.R. Towler, The role of poly(acrylic acid) in conventional glass polyalkenoate cements. J. Polym. Eng. **36**(3), 221–237 (2015). https://doi.org/10.1515/POLYENG-2015-0079/ASSET/GRAPHIC/J_POLYENG-2015-0079_FIG_005.JPG

40. DePuySynthes, *Depuy CMW Orthopaedic Bone Cements—Frequently Asked Questions.* DePuy Synthes joint reconstruction, 2014. Accessed 10 Dec 2022. [Online]. Available: http://synthes.vo.llnwd.net/o16/LLNWMB8/INT%20Mobile/Synthes%20International/ Product%20Support%20Material/legacy_Synthes_PDF/DPEM-BIOM-1213-0004_LR.pdf
41. Y. Chang, C.-L. Tai, P.-H. Hsieh, S.W.N. Ueng, Gentamicin in bone cement. Bone Joint Res. **2**(10), 220–226 (2013). https://doi.org/10.1302/2046-3758.210.2000188
42. GEAdditive, *3D-Printed Joints & Implants: 100,000 Patients Later, The 3D-Printed Hip Is A Decade Old And Going Strong.* GE Reports. Accessed 06 April 2024. [Online]. Available: https://www.ge.com/additive/ja/node/1199
43. S. Saunders, *GE Reports Looks to The Future While Celebrating 10th Anniversary of First 3D Printed Hip Implant.* 3DR Holdings. Accessed 06 April 2024. [Online]. Available: https://3dp rint.com/206019/ge-reports-3d-printed-hip-implant/
44. P.D. Olson, *100,000 Patients Later, the 3D-Printed Hip Is a Decade Old and Going Strong.* GE Reports. Accessed 06 April 2024. [Online]. Available: https://www.odtmag.com/contents/ view_breaking-news/2018-03-05/100000-patients-later-the-3d-printed-hip-is-a-decade-old-and-going-strong/
45. Z. Gaertner, *3D Printing & Medicine.* Medium. Accessed 06 April 2024. [Online]. Available: https://medium.com/@gaer9121/3d-printing-medicine-d5a752b9d8ea
46. V. Krebs, *3D-Printed Implant Reconstructs Hip in Patient with No Proximal Femur or Pelvis.* Cleveland Clinic. Accessed 06 April 2024. [Online]. Available: https://consultqd.clevelandcli nic.org/3d-printed-implant-reconstructs-hip-in-patient-with-no-proximal-femur-or-pelvis
47. US_FDA, File:pelvic component for hip joint replacement (5185) (18494294741).jpg. *Wikimedia Commons.* Accessed 24 June 2024. [Online]. Available: https://commons.wikimedia. org/wiki/File:Pelvic_Component_for_Hip_Joint_Replacement_(5185)_(18494294741).jpg
48. K. Čolić, A. Sedmak, A. Grbovic, U. Tatić, S. Sedmak, B. Đorđević, Finite element modeling of hip implant static loading. Procedia Eng. **149**, 257–262 (2016). https://doi.org/10.1016/j.pro eng.2016.06.664
49. K. Čolić et al., Eksperimentalno i numeričko istraživanje mehaničkog ponašanja umjetnog kuka od legure titana [Experimental and numerical research of mechanical titanium alloy of hip implant]. Tehnicki Vjesnik [Technical Gazette] **24**(3), 709–713 (2017). https://doi.org/10. 17559/TV-20160219132016
50. Suyitno, L. Sutomo, Static test and simulation of hip joint prosthesis. In: *2017 7th International Annual Engineering Seminar (InAES).* (IEEE, 2017), pp. 1–4. https://doi.org/10.1109/INAES. 2017.8068566
51. S. Arabnejad, B. Johnston, M. Tanzer, D. Pasini, Fully porous 3D printed titanium femoral stem to reduce stress-shielding following total hip arthroplasty. J. Orthop. Res. **35**(8), 1774–1783 (2016). https://doi.org/10.1002/jor.23445
52. J. Corona-Castuera, D. Rodriguez-Delgado, J. Henao, J.C. Castro-Sandoval, C.A. Poblano-Salas, Design and fabrication of a customized partial hip prosthesis employing CT-scan data and lattice porous structures. ACS Omega **6**(10), 6902–6913 (2021). https://doi.org/10.1021/ acsomega.0c06144
53. M.A. Ahmad, N.N.M.E. Zulkifli, S. Shuib, S.H. Sulaiman, A.H. Abdullah, Finite element analysis of proximal cement fixation in total hip arthroplasty. Int. J. Technol. **11**(5), 1046 (2020). https://doi.org/10.14716/ijtech.v11i5.4318
54. S. Tabaković, M. Zeljković, Z. Milojević, A. Živković, Design of custom made prosthesis of the hip. Technical Gazette **26**(2), 323–330 (2019). https://doi.org/10.17559/TV-20171006104842
55. A. Pacioga, D.D. Palade, S. Comşa, Computational Simulation of bone-personalized hip prosthesis assembly. UPB Sci. Bull., Ser. D: Mech. Eng. **73**(2), 2429–262 (2011) Accessed 27 Jan 2022. [Online]. Available: https://www.scientificbulletin.upb.ro/rev_docs_arhiva/full15 939.pdf
56. Y. Jun, K. Choi, Design of patient-specific hip implants based on the 3D geometry of the human femur. Adv. Eng. Softw. **41**(4), 537–547 (2010). https://doi.org/10.1016/j.advengsoft. 2009.10.016

Chapter 3
Pugh Chart Method and Finite Element Analysis (FEA) of Hip Implant

3.1 Introduction

The Pugh chart method is used to choose which of the existing hip prosthesis is the best choice to undergo a redesign. It is a qualitative technique used to compare and select the best option from a set of alternatives based on multiple criteria. Each criterion is assigned a score, and the alternative that scores the highest overall is considered the best choice [1]. It is particularly useful when making decisions that involve complex trade-offs between different factors. Table 3.1 provides a summary of the characteristics of existing hip prostheses between Charnley and 3M Capital for design selection, prior to the evaluation using the Pugh chart method.

3.1.1 Bone Ingrowth in Hip Prosthesis

Bone ingrowth, also known as osseointegration, refers to the process by which living bone tissue grows into the surface of an implant, creating a strong and stable connection between the implant and the surrounding bone, helping to prevent implant loosening and improving the overall function and longevity of the prosthetic joint [7]. This phenomenon is crucial for the long-term success and stability of orthopaedic implants, such as hip prosthesis.

Porous coatings or surfaces on hip prosthesis play a significant role in promoting bone ingrowth. These porous structures encourage bone tissue to grow into the implant, enhancing mechanical stability and reducing the risk of implant migration or loosening [8]. Additionally, porous coatings increase the surface area available for bone attachment, facilitating faster and more extensive osseointegration.

For this research, a 5 mm length slot is added at the neck of the hip prosthesis with different radius of 1 mm and 2 mm, respectively, to promote the bone in growth in the modified design.

N. S. Hamizan et al., *Hip Prosthesis*,
SpringerBriefs in Applied Sciences and Technology,
https://doi.org/10.1007/978-981-96-1470-7_3

Table 3.1 Characteristics between existing hip prosthesis [2–6]

Design	Charnley		3M Capital	
	Roundback (2nd gen.)	Flanged (3rd gen.)	Roundback	Flanged
Image[1]				
Characteristics	• Rounded cross-sectional profile • Tapered geometry • Asymmetrical cross	• Mini-internal collar • Tapered geometry • Dorsal flange	• More metal was added to the flat lateral surface of the stem • Cross-section not entirely oval or round	• Rounded medial and flange added to the lateral surfaces below the shoulder • Irregular cross-section
Surface finish	• Rough surface	• Polished smooth surface	• Matt surface	• Rough surface
Advantages	• Good mechanical stability • Biocompatibility with bone • Enhanced load transfer to cement	• Longer life span • Good mechanical and torsional stability within the cement mantle • Biocompatibility with bone	• Better survival • Better rotation stability • Resistance to oxidation	• High mechanical strength and corrosion resistance • Biocompatibility with bone
Disadvantages	• Decreased resistance to wear	• Decreased resistance to wear	• Poor biocompatibility with bone • Low aseptic loosening rates	• Decreased resistance to wear • High aseptic loosening

[1] Used with permission of SLACK Incorporated, from Philosophies of stem designs in cemented total hip replacement, Nico Verdonschot, 28, 8 Suppl and 2005; permission conveyed through Copyright Clearance Center, Inc.

3.2   Finite Element Analysis (FEA)

FEA is a problem-solving method widely used to solve engineering and mathematical models. It involves utilising a numerical technique referred to as the finite element method (FEM) to simulate and analyse various physical phenomena in simple or complex geometries. FEA allows for virtual experiments that incorporate real-world properties, making it possible to optimise designs using specialised simulation software [9–11] allowing engineers and designers to accelerate the product development process and achieve better results.

There are several notable simulation software options available to conduct FEA, such as SolidWorks, ANSYS, Autodesk, ABAQUS, etc. These software packages provide comprehensive tools and capabilities to perform detailed simulations, enabling users to identify potential issues in designs. Through FEA, one can precisely locate and address areas of tension, weak spots, or other structural concerns without the need to produce physical prototypes. This capability not only is cost-saving to the production line but also streamlines the design iteration process.

There are some pre-processing and post-processing steps when conducting FEA. The flowchart can be seen in Fig. 3.1. The pre-processing method includes in creating or importing CAD model, selecting material and its properties, generation of mesh, and applying necessary boundary conditions and loads, whereas the post-processing method provides the analysis data and the visualisation of the computed results.

Static structural analysis was performed to run the simulation. The femoral stem used is the Charnley stem, with its parts shown in Fig. 3.2, and the dimensions of the Charnley hip stem prosthesis are tabulated in Table 3.2.

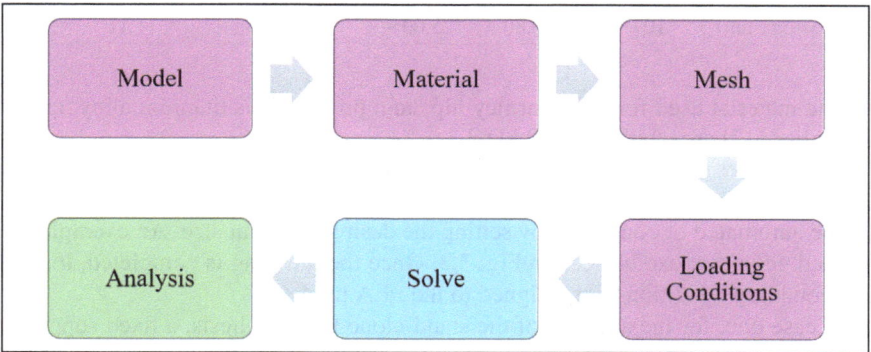

Fig. 3.1 Flowchart of FEA stages: pre-processing in purple, processing in blue and post-processing in green

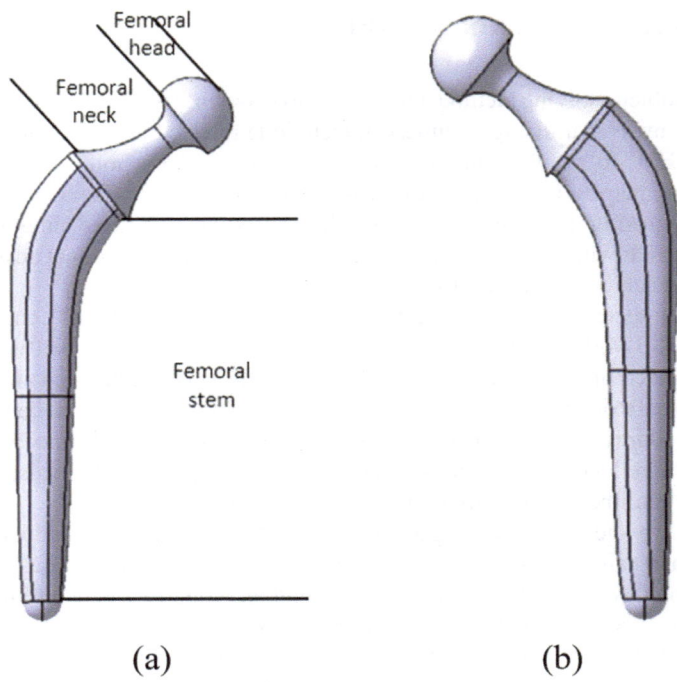

Femoral head

Femoral neck

Femoral stem

(a)   (b)

Fig. 3.2 Charnley hip prosthesis: **a** Parts **b** before modification [2]

Table 3.2 Dimensions of the Charnley hip prosthesis

Part	Length		Diameter		Radius
	Stem	Neck	Neck-stem	Neck-head	Femoral head
Dimension (mm)	107.70	28.80	25.00	8.69	11.11

The material used for the Charnley hip stem prosthesis is titanium alloy, and its properties [12] are tabulated in Table 3.3.

FEA divides solid geometry into smaller discrete elements [13] through a meshing process, allowing for a detailed representation of the structure. This meshing process can be automated or controlled by setting the desired element size. An example of a meshed geometry can be seen in Fig. 3.3. Once the meshing is completed, loading and boundary conditions are assigned to the FEA model.

In case one, for the analysis of the stand-alone hip prosthesis, a fixed support is applied to the lower region of the hip prosthesis, simulating its fixation within the bone and a head load force acting at the femoral head, representing load during various activities. Meanwhile, for case two of the hip implant-PMMA-bone assembly, the

[2] Reprinted from N. S. Hamizan et al., "Design for additive manufacturing (DfAM) for hip prosthesis for 3D printing," AIP Conf Proc, vol. 2571, p. 030005, 2023, doi: https://doi.org/10.1063/5.0115923, with the permission of AIP Publishing.

Table 3.3 Material properties of the bone assembly

Material	Poisson's ratio, v	Density, ρ	Young's modulus, E	Tensile	Compression
		kg/m^3	MPa	MPa	MPa
Bone (Femur)	0.39	(vary by several factors)*	17,200	121	167
PMMA	0.36	1,188	4,500	21	144
Ti–6Al–4V	0.34	4,470	110,000	965	960

*The asterisk highlights that bone density is not a fixed value and can be influenced by various factors, such as the individual's gender, age, physical activity level, hormonal changes, nutrition, lifestyle choices, medical conditions, and genetic predisposition. This note provides context for interpreting any data or values related to bone density in the table

Fig. 3.3 Mesh generated on the existing Charnley hip prosthesis [3]

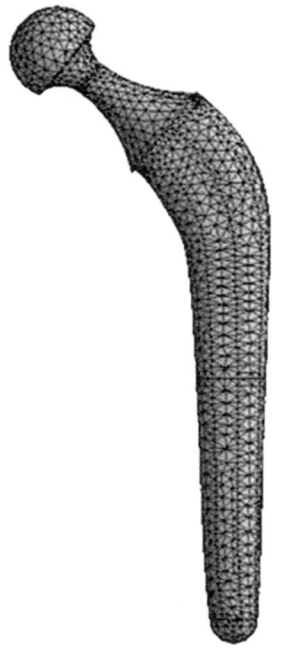

distal end of the bone was secured in place, and additionally, the head and abductor load forces were applied to the hip prosthesis.

Table 3.4 shows the head and abductor load forces acting on the hip prosthesis, which reflect or mimic the real-world scenarios experienced by the hip joint, that are extracted from previous studies [14–16]. For this research, however, only one load case was considered which is the standing position.

---

[3] Reprinted from N. S. Hamizan et al., "Design for additive manufacturing (DfAM) for hip prosthesis for 3D printing," AIP Conf Proc, vol. 2571, p. 030005, 2023, doi: https://doi.org/10.1063/5.0115923, with the permission of AIP Publishing.

Table 3.4 Head and abductor load acting on hip prosthesis

Types of loading	F_x (N)	F_y (N)	F_z (N)	Head load, F (N)
				Abductor load, F (N)
Standing	1,744.70	0.00	−927.68	1,976.00
	−949.89	0.00	797.40	1,240.00
Walking	2,029.58	382.30	−339.47	2,093.00
	−798.70	382.30	−551.62	972.89
Stair climbing	2,087.40	639.55	378.97	2,215.80
	−832.02	639.55	−686.98	1,115.30

Next, analysis is done on the FEA models. For this research, the focus is on analysing the total deformation, equivalent (von Mises) stress and shear stress within the hip prosthesis, femoral bone, and bone cement. The results of the FEA simulation are usually illustrated in a colour scale ranging from maximum to minimum values of the analysis done. This visual representation allows for easy interpretation of deformation and stress, highlighting regions of high or low-stress concentrations. The analysis values are later compared to the compressive load testing that will be conducted after the fabrication of the hip prosthesis and jig.

3.2.1 Stress Distribution

The stress distributions within the bone-implant interface play a crucial role in determining the long-term performance of a hip prosthesis. It is primarily influenced by the effective area of the implant, which directly affects the way stress is distributed within the surrounding bone [17]. Stress, being a function of load and area, means that any changes in the effective area of the interface may affect the stress distribution to the bone.

A mechanical phenomenon that is associated with the stress distribution in the bone is stress shielding. It occurs when the implant absorbs a portion of the load that would otherwise be transferred to the surrounding bone [18]. This reduction in load transfer can lead to bone resorption and subsequently weaken the bone structure. As a result, the stress distributed at the bone-implant interfaces decreases, leading stress shielding to occur.

Shielding of the bone from stress results in bone adaptation which involves metabolic changes with internal or external remodelling, resulting in bone mass reduction. The bone may become more porous or thinner as a natural adaptation process corresponding to the decreased carrying load of the bone. However, this bone structure weakening can lead to mechanical failure, including bone loss or bone resorption, and ultimately result in stem loosening at the implant-bone interface [19]. Consequently, when performing the THA procedure, it is important to consider maintaining the pre-operative load transfer and stress distribution to the bone to minimise the risk of bone resorption for long-term success and stability of the hip implant.

The present invention aims to address the issue of the stress shielding associated with existing implant design. It is expected that the design of the hip implant will result in lower stress distribution and reduce the occurrence of stress shielding. The closer the stress distribution resembles the original bone, the better the hip implant is in terms of preserving the bone's integrity and minimising the risk of complications. Therefore, the invention seeks to evaluate the long-term performance of hip prosthesis by optimising the stress distribution and minimising the stress shielding effects.

3.2.2 Total Deformation (mm)

Total deformation refers to the overall displacement or distortion of a structure or component under applied loads. FEA is a numerical method used to analyse the behaviour of structures and systems by dividing them into smaller FE and solving equations to simulate their response to various conditions.

Total deformation is a result of the combined effects of both elastic deformation and plastic deformation. Elastic deformation occurs when a material undergoes reversible changes in shape or size due to applied loads. It is characterised by temporary deformation, meaning that once the load is removed, the material returns to its original shape. Plastic deformation, on the other hand, involves permanent changes in the shape or size of a material beyond its elastic limit. This occurs when the applied load exceeds the material's yield strength, causing the material to undergo permanent deformation [20]. Equation (3.1) shows the equation of deformation:

$$\Delta L = \frac{1}{E} \frac{F}{A} L_0 \tag{3.1}$$

where ΔL is the amount of deformation or change in length, E is Young's modulus, F is the applied force, A is the cross-sectional area, and L_0 is the original or initial length [21].

3.2.3 Equivalent (von Mises) Stress (MPa)

Equivalent (von Mises) stress is a commonly used stress measure to assess the structural integrity and failure potential of components or structures. It is a scalar value that represents the combined effects of normal and shear stresses on a material [22]. It is particularly useful in situations where materials may exhibit different strengths under tension and compression. By considering the combined stress state, equivalent (von Mises) stress provides a measure of the critical stress level that can cause yielding or failure in a ductile material [22]. If the equivalent (von Mises) stress exceeds the yield strength of the material, it indicates that yielding or plastic deformation is likely to occur in that region. The equivalent (von Mises) stress equation for a three-dimensional state of stress can be expressed as in Eq. (3.2):

$$\begin{bmatrix} \sigma_1 & 0 & 0 \\ 0 & \sigma_2 & 0 \\ 0 & 0 & \sigma_3 \end{bmatrix} \quad \sigma_{vm} = \sqrt{\frac{(\sigma_1 - \sigma_2)^2 + (\sigma_2 - \sigma_3)^2 + (\sigma_3 - \sigma_1)^2}{2}} \qquad (3.2)$$

where σ_{vm} is the equivalent (von Mises) stress and σ_1, σ_2 and σ_3 are the principal stresses [22, 23].

3.2.4   Shear Stress (MPa)

Shear stress is the distribution of internal forces within a structure or component. It represents the force per unit area acting parallel to a specific plane, causing one layer of the material to slide or deform relative to an adjacent layer [24]. It is derived from the shear force and the cross-sectional area of the material. It is a measure of the intensity of internal forces that cause shearing deformation in the material. The general equation of shear stress is defined as in Equation (3.3):

$$\tau = \frac{F}{A} \qquad (3.3)$$

where τ is the shear stress, F is the force applied parallel to the surface and A is the cross-sectional area of the material through which the force is applied [24].

3.3   Preparation of Jig and Testing of Hip Prosthesis

Prior to the compressive load testing of the hip prosthesis, several preparation steps are required to prepare for it. First and foremost, as there are various international standards that are related to the testing of hip prosthesis, the appropriate standard is to be determined for the benchmarking process. Then, the fabrication of the intended hip prosthesis design is done via 3D printing. Finally, preparation of PMMA is also needed before undergoing the compressive load testing.

3.3.1   Compressive Load Testing of Hip Prosthesis

There are several internationally recognised standards to test a hip prosthesis. The ISO 7206 Standards serve as the benchmark for conducting these tests and ensuring the quality and reliability of hip prosthesis. This standard consists of several components,

Fig. 3.4 Illustration of a
simple compression testing
process done on a hip
prosthesis [30][4]

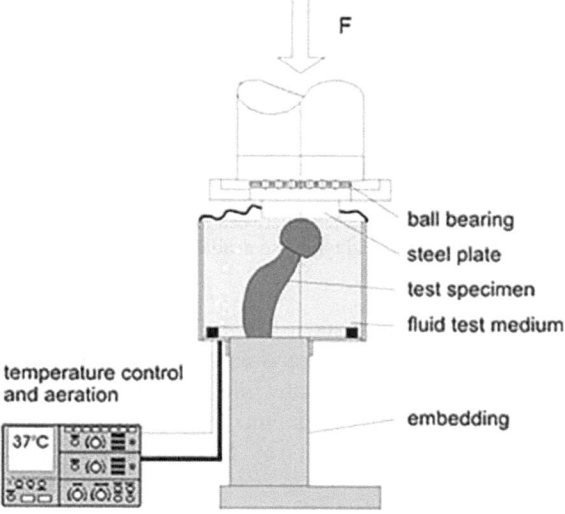

such as ISO 7206-4 (2010) Implants for surgery—Partial and total hip joint prostheses—Part 4: Determination of endurance properties and performance of stemmed
femoral components [25, 26], ISO 7206-6 (2013) Implants for surgery—Partial and
total hip joint prostheses—Part 6: Endurance properties testing and performance
requirements of the neck region of stemmed femoral components [27], and ISO
7206-10 (2018) Implants for surgery—Partial and total hip joint prostheses—Part
10: Determination of resistance to static load of modular femoral heads [28, 29].

However, for this research, the testing of the hip prosthesis is only focusing on
the ISO 7206-4 Standard [25, 26] to assess the durability and overall functionality of
stemmed femoral components within the hip prosthesis. The fatigue test according
to this standard simulates the dynamic loading experienced by a hip stem during gait
condition. An example of a fatigue test done on a hip prosthesis is shown in Fig. 3.4.

To simulate the worst-case scenarios and replicate clinical failure, the test focuses
on the proximal loosening of the stem. By subjecting the hip prosthesis to these conditions, the test provides valuable insights into its resilience and ability to withstand
demanding loads. In addition to that, this standard also specifies various parameters
that govern the fatigue test, including load inclination, embedding level, load level,
and a number of cycles to be applied for the test [30]. Nevertheless, it is important
to note that for this research, only static compressive loading will be applied.

The compressive load testing was conducted on the hip prosthesis by using a
Universal Testing Machine (UTM) called Shimadzu Servopulser E-type loading
frame testing machine equipped with a Servo Controller 4830. The testing machine is
available at the Strength of Materials Laboratory, School of Mechanical Engineering,
College of Engineering, Universiti Teknologi MARA.

---

[4] Reproduced with permission from Endolab Mechanical Engineering GmbH.

Table 3.5 Major specifications of Shimadzu Servopulser E-type loading frame testing machine used [31]

Main unit model	Stroke	Maximum test force Dynamic/Static	Frame rigidity[*1]	Crosshead drive mechanism[*2]
E101 kN	±25 mm	±100 kN/±120 kN	0.0012 mm/kN	Hydraulic drive with hydraulic clamp

* Note1 At 500 mm clearance between crosshead and table
* Note2 Fixed crosshead type also available (without drive mechanism or hydraulic clamp)

The UTM has a standard frame design featuring a bottom actuator and can cater for the testing of small parts and formed specimens, allowing for load tests of up to force of 200 kN. Meanwhile, Controller 4830 is a compact controller with a colour LCD and convenient touch panel interface that can generate a comprehensive variety of loading waveforms. It can handle the measurement, control and display the waveforms.

The UTM possesses the capability to conduct tests such as tension, compression, bending and torsion. The standard frame type requires a small installation space and is suited for testing formed specimens and small parts. The highly rigid frame machine prevents the buckling of specimens and minimises losses in hydraulic energy caused by frame deformation [31]. Moreover, the crosshead hydraulic drive mechanism simplifies crosshead vertical movement and clamping. Additionally, the machine allows for the attachment of various test jigs and environment control devices. Major test applications include the evaluation of metal and plastic specimens, fracture toughness testing, as well as testing of standard specimens and small parts [31]. The major specification of the machine can be referred to in Table 3.5.

3.3.2 Implant Prototyping Using Additive Manufacturing (AM) Technology

In line with IR 4.0, AM has become one of the most advanced technologies and manufacturing in these recent years. Through the usage of 3D printers, AM is the science, art and skill to design for manufacturability [32]. Through this technology, very complex shapes can be produced, without the need to assemble as all parts are printed in place [33] while minimising the weight and material consumption, different from traditional manufacturing [32]. Problems can be identified and solved on the spot during the assembly line, thus allowing for rapid and mass prototyping of a better product. Easy and instant sharing of digital files enables vast collaborations in numerous locations due to the quick access using CAD models [32].

Table 3.6 shows the differences between conventional manufacturing and AM extracted from [34–36].

Table 3.6 Comparison between conventional manufacturing and AM

Conventional/Traditional manufacturing	AM/3D printing/Rapid prototyping
Involves removing parts of materials to create the desired shape	Operates by adding layer-by-layer of materials to create product
Parts are produced separately before assembling	Can combine FEA with topology optimisation
Difficult or impossible to create complex or customised parts	Enable mass production of complex, customised or fully assembly product
Products are usually heavier	Products are usually stronger and lighter
High material and production costs, time-consuming	Low material and production cost, low production time
Requires several steps using different machines and post-processing	Less labour and manufacturing methods
More material waste	Minimal material waste

3.4 Selective Laser Melting

Titanium is expensive material and can be a challenge when it comes to conventional processing technologies. The relatively low-cost precision of 3D printing is therefore a game-changer for titanium [37].

Selective laser melting (SLM) is a high-performance metal AM technique that selectively melts a metal powder bed, track by track and layer by layer to construct a 3D metal part [38, 39], as illustrated in Fig. 3.5. It is an AM method specially developed for 3D printing metal alloys. It creates parts additively by fusing metal powder particles together in a full melting process, allowing the metal to form a homogeneous block with good resistance. It fits perfectly with pure metals like titanium or aluminium [40].

SLM process uses supports to reinforce small angles and hangovers of the parts but also to stick the design to the job tray. The support will be removed manually after cooling. Various finishing techniques, such as milling or heat treatment, are commonly used to achieve the functional requirements of the part [40]. At present, the SLM technology has been utilised in many industrial fields, for instance, aerospace [40, 42, 43], automotive [44, 45], dental [46, 47] and medical engineering [38, 48, 49].

Compared to other high-performance metal AM, the SLM is claimed to have the advantages of fabricating parts with a complex shape and high dimensional accuracy [50]. Several advantages of SLM include allowing the creation of complex geometries and functional metal structures to get lightweight designs, providing productivity and cost-effective products, as compared to conventional manufacturing processes such as casting, forging and machining.

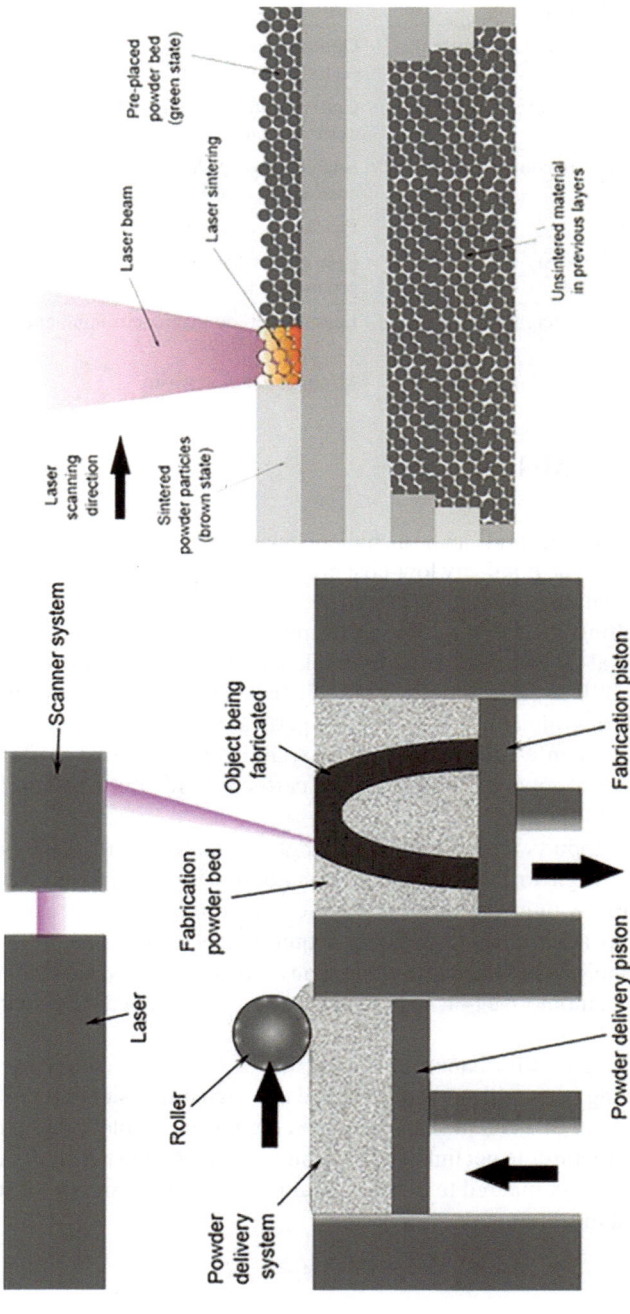

Fig. 3.5 SLM system schematic [41][5]

3.4.1 PMMA Preparation

In general, a package of 40 g PMMA units came with two sterilised components: liquid and powder. The liquid component consists of flammable substance, colourless in appearance and has a prominent smell. The powder component is a white, finely divided powder consisting of a PMMA-based polymer.

Bone cement with medium viscosity (MV) or high viscosity (HV) is available on the market. HV bone cement offers a faster pick-up time compared to MV bone cement, and the working time is geared towards cementing procedures that require a longer working time, such as a fully cemented total knee replacement.

Manufacturer's instructions on PMMA preparation are pretty much like each other. Bone cement is sensitive to heat. Therefore, any increase or decrease in temperature, either in the ambient temperature and/or of the cement components and mixing equipment, from the recommended temperature of 23 °C will affect the handling characteristics and setting time of the cement [51, 52]. In addition to that, the variations in humidity will affect the cement handling characteristics and the setting time of the cement. Moreover, vacuum mixing under vacuum accelerates the handling times of PMMA as compared to manual mixing. For proper fixation, it is necessary to maintain the positioning of the prosthetic or prosthesis components until the completion of the polymerisation process upon application of the bone cement [51, 52].

Bone cement is also available with gentamicin admixture. Gentamicin bone cement is antibiotic loaded with 1 g of active gentamicin in a standard 40 g package of bone cement. This gentamicin is effective in reducing the risks posed by gram-positive and a few gram-negative bacteria [51, 52]. Furthermore, antibiotic-loaded bone cement has been shown to reduce the risk of infection, as the antibiotic in the cement provides effective prophylaxis against infection, resulting in an incidence of deep infection of only 1.2%, compared with 2.3% when no antibiotic was used in the bone cement. Bone cement with added gentamicin results in much higher levels of gentamicin in the joint than can be achieved with systemic antibiotics [51, 52].

Some precautions in preparing the PMMA include that the mixture should be thoroughly and carefully mixed and rotated in one direction only, either clockwise or anticlockwise, to avoid any air bubbles being trapped inside the PMMA mixture. It is also advisable to wear a face mask as the PMMA exhibits a very strong odour during mixing. Besides that, insertion of the prosthesis should be done at an appropriate time, at sufficient viscosity to prevent excessive displacement of the prosthesis, but also should not be so long that there is a risk that the procedure cannot be completed due to hardening of the cement. The prosthesis must then be held firmly in place to prevent movement, and pressure must be maintained until the cement has fully cured. Excess bone cement must be removed prior to the hardening process.

For the use of PMMA in surgery, it is important to store the PMMA packaging at the recommended mixing temperature (23 °C) for at least 24 hours before use. Additionally, the packaging is guaranteed sterile if it is unopened or undamaged. Likewise, gentamicin bone cement is intended for single use only and cannot be reused or re-sterilised [51, 52].

3.5 Summary

This chapter summarises Pugh chart method, FEA, preparation of jig and compressive load testing of the hip prosthesis. Chapter 4 will introduce the approach of this research which includes design selection, FEA on stand-alone hip prosthesis and bone assembly with and without PMMA, fabrication of jig and 3D printed modified hip prosthesis, preparation of PMMA and compressive load testing of the hip prosthesis.

References

1. T. Hessing, *Pugh Analysis, Six Sigma Study Guide.* Accessed 06 Oct 2023. [Online]. Available: https://sixsigmastudyguide.com/pugh-analysis/
2. *Royal College of Surgeons of England.* An investigation of the performance of the 3M Capital hip system: July 2001. (2001)
3. D. Janssen, R. Aquarius, J. Stolk, N. Verdonschot, Finite-element analysis of failure of the Capital Hip designs. J. Bone Joint Surg. Br. **87**-B(11), 1561–1567 (2005). https://doi.org/10.1302/0301-620X.87B11.16358
4. L.A. Reynolds, E.M. Tansey (Eds.) *Early Development Of Total Hip Replacement*, vol. 29. London: Wellcome Trust Centre for the History of Medicine at UCL, 2007. Accessed 24 Oct 2020. [Online]. Available: http://www.histmodbiomed.org/sites/default/files/44852.pdf
5. N. Verdonschot, Implant choice: stem design philosophies. In: *The Well-Cemented Total Hip Arthroplasty*, 1st ed., (Springer, Berlin, 2005). ch. 7.1, pp. 168–179. https://doi.org/10.1007/3-540-28924-0_22
6. Healio, *Evolution of Cemented Stems.* OrthopedicsToday. Accessed 24 Oct 2020. [Online]. Available: https://www.healio.com/news/orthopedics/20120331/evolution-of-cemented-stems
7. S. Parithimarkalaignan, T.V. Padmanabhan, Osseointegration: an update. J. Indian Prosthodont. Soc. **13**(1), 6 (2013). https://doi.org/10.1007/S13191-013-0252-Z
8. A. Bandyopadhyay, I. Mitra, J.D. Avila, M. Upadhyayula, S. Bose, Porous metal implants: processing, properties, and challenges. Int. J. Extreme Manuf. **5**(3), 032014 (2023). https://doi.org/10.1088/2631-7990/acdd35
9. T. English, What is finite element analysis and how does it work?. Interesting Eng. Accessed 29 Oct 2020. [Online]. Available: https://interestingengineering.com/what-is-finite-element-analysis-and-how-does-it-work
10. What is FEA | finite element analysis?. *SimScale Documentation.* Accessed 29 Oct 2020. [Online]. Available: https://www.simscale.com/docs/simwiki/fea-finite-element-analysis/what-is-fea-finite-element-analysis/
11. What is finite element analysis (FEA)?. TWI. Accessed 29 Oct 2020. [Online]. Available: https://www.twi-global.com/technical-knowledge/faqs/finite-element-analysis
12. K.S. Katti, Biomaterials in total joint replacement. Colloids Surf. B **39**(3), 133–142 (2004). https://doi.org/10.1016/j.colsurfb.2003.12.002
13. R. Kulshrestha, Finite element analysis (FEA)—an insight. Mathews J. Dent. **3**(2), 22 (2018). Accessed 03 Nov 2020. [Online]. Available: https://www.mathewsopenaccess.com/scholarly-articles/finite-element-analysis-f-e-a-an-insight.pdf
14. S. Shuib, N.F. Ismail, M.A. Yahya, A.A. Shokri, Analysis of an improved hybrid stem design for total hip replacement (THR). J. Mech. Eng. **5**(5), 205–215 (2018)
15. S. Shuib, B.B. Sahari, W.S. Voon, M. Arumugam, Finite elemental analysis of outer and inner surfaces of the proximal half of an intact femur. Trends Biomater. Artif. Organs **26**(2), 103–106 (2012) Accessed 27 July 2020. [Online]. Available: https://www.researchgate.net/publication/260593087_Finite_elemental_analysis_of_outer_and_inner_surfaces_of_the_proximal_half_of_an_intact_femur
16. M. Viceconti, L. Bellingeri, L. Cristofolini, A. Toni, A comparative study on different methods of automatic mesh generation of human femurs. Med. Eng. Phys. **20**(1), 1–10 (1998). https://doi.org/10.1016/S1350-4533(97)00049-0

17. J.M. Jung, C.S. Kim, Analysis of stress distribution around total hip stems custom-designed for the standardized Asian femur configuration. Biotechnol. Biotechnol. Equip. **28**(3), 525–532 (2014). https://doi.org/10.1080/13102818.2014.928450

18. S.A. Naghavi et al., A novel hybrid design and modelling of a customised graded Ti–6Al–4V porous hip implant to reduce stress-shielding: an experimental and numerical analysis. Front Bioeng Biotechnol **11**, 1–20 (2023). https://doi.org/10.3389/fbioe.2023.1092361

19. T.D. Szwedowski, W.R. Taylor, M.O. Heller, C. Perka, M. Müller, G.N. Duda, Generic rules of mechano-regulation combined with subject specific loading conditions can explain bone adaptation after THA. PLoS ONE **7**(5), e36231 (2012). https://doi.org/10.1371/journal.pone. 0036231

20. FasterCapital, Deformation: beyond the surface: exploring elastic deformation in solids. *Faster Capital*. Accessed 06 Nov 2023. [Online]. Available: https://fastercapital.com/content/Deform ation--Beyond-the-Surface--Exploring-Elastic-Deformation-in-Solids.html

21. OpenStax, 71 10.1 force-deformation curve. *Biomechanics of Human Movement*. Accessed 06 Oct 2023. [Online]. Available: https://pressbooks.bccampus.ca/humanbiomechanics/ chapter/4-2-hookes-law-originally-section-5-3-elasticity-stress-and-strain-2/#:~:text=HOO KE'SLAW-,F%3DkΔL%2C, the direction of the force

22. S. Das, *Von Mises Stress Versus Principal Stress: Analysis for Engineers*. Lambda Geeks. Accessed 06 Oct 2023. [Online]. Available: https://lambdageeks.com/von-mises-stress-vs-pri ncipal-stress/?msclkid=b196ffbeaed611ecbeb81d0a23dd8f64

23. ANSYS, *What is Equivalent Stress?*. ANSYS BLOG. [Online]. Available: https://www.ansys. com/blog/what-is-equivalent-stress

24. Xometry, *Shear Stress: Definition, How it Works, Example, and Advantages*. Xometry. Accessed 06 Oct 2023. [Online]. Available: https://www.xometry.com/resources/materials/shear-stress/

25. ISO, ISO 7206-4:2010—implants for surgery—partial and total hip joint prostheses—part 4 determination of endurance properties and performance of stemmed femoral components. (International Organization for Standardization, 2010)

26. ISO, ISO 7206-4:2010 AMD 1:2016-ISO 7206-4:2010—implants for surgery—partial and total hip joint prostheses—part 4 determination of endurance properties and performance of stemmed femoral components—amendment 1. (International Organization for Standardization, 2016)

27. ISO, ISO 7206-6:2013—implants for surgery—partial and total hip joint prostheses—part 6 endurance properties testing and performance requirements of neck region of stemmed femoral components. (International Organization for Standardization, 2013)

28. ISO, ISO 7206-10:2018—implants for surgery—partial and total hip-joint prostheses—part 10 determination of resistance to static load of modular femoral heads. (International Organization for Standardization, 2018)

29. ISO, ISO 7206-10:2018 AMD 1:2021—implants for surgery—partial and total hip-joint pros-theses—part 10 determination of resistance to static load of modular femoral heads—amend-ment 1. (International Organization for Standardization, 2021)

30. Hip joint prostheses, *EndoLab Mechanical Engineering GmbH*. Accessed 13 Nov 2020. [Online]. Available: https://www.endolab.org/implant-testing.asp?cat1=1&topic=Hipjointp rostheses&desc=Hipimplantteststandards/Hip-Implantstesting&key=hip,implant,implanttesti ng,certified,accredited,fatiguetesting,statictesting,weartesting,luxation,pull-of

31. EHF-E series: servopulser fatigue and endurance testing machine. *Shimadzu Corporation* (2021). Accessed 20 May 2022. [Online]. Available: https://www.shimadzu.com/an/products/ materials-testing/fatigue-testingimpact-testing/ehf-e-series/index.html

32. M. Torosian, What is design for additive manufacturing?. *Jabil Additive*. Accessed 24 Oct 2020. [Online]. Available: https://www.jabil.com/blog/design-for-additive-manufacturing.html

33. P. Keane, What is design for additive manufacturing?—engineers rule. *Engineers Rule*. Accessed 24 Oct 2020. [Online]. Available: https://www.engineersrule.com/design-additive-manufacturing/

34. T. Pereira, J.V. Kennedy, J. Potgieter, A comparison of traditional manufacturing versus additive manufacturing, the best method for the job. Procedia Manuf. **30**, 11–18 (2019). https://doi.org/ 10.1016/j.promfg.2019.02.003

35. B. Warburg, Additive manufacturing versus traditional manufacturing—animal ventures. *Animal Ventures*. Accessed 29 Oct 2020. [Online]. Available: https://blog.animalventures.com/blog/additive-manufacturing-vs-traditional-manufacturing/
36. J.W. Booth, J. Alperovich, P. Chawla, J. Ma, T.N. Reid, K. Ramani, The design for additive manufacturing worksheet. J. Mech. Des. **139**(10), 100904 (2017). https://doi.org/10.1115/1.4037251
37. Parts and materials for hip replacement. *Hip and Knee*. Accessed 27 Jan 2022. [Online]. Available: https://hipandknee.com/hip-surgery/about-hip-replacement/parts-materials/
38. F. Bartolomeu et al., Multi-material Ti–6Al–4V & PEEK cellular structures produced by selective laser melting and hot pressing: a tribocorrosion study targeting orthopedic applications. J. Mech. Behav. Biomed. Mater. **89**, 54–64 (2019). https://doi.org/10.1016/j.jmbbm.2018.09.009
39. L. Zhang, S. Zhang, H. Zhu, Z. Hu, G. Wang, X. Zeng, Horizontal dimensional accuracy prediction of selective laser melting. Mater. Des. **160**, 9–20 (2018). https://doi.org/10.1016/j.matdes.2018.08.059
40. 3D printing with SLM technology. *Sculpteo*. Accessed 30 Nov 2020. [Online]. Available: https://www.sculpteo.com/en/materials/slm-material/
41. Materialgeeza, File:selective laser melting system schematic.jpg. *Wikimedia Commons*. Accessed 24 Oct 2020. [Online]. Available: https://commons.wikimedia.org/wiki/File:Selective_laser_melting_system_schematic.jpg
42. M. Seabra et al., Selective laser melting (SLM) and topology optimization for lighter aerospace componentes. Procedia Struct. Integrity **1**, 289–296 (2016). https://doi.org/10.1016/j.prostr.2016.02.039
43. M. Brandt, S.J. Sun, M. Leary, S. Feih, J. Elambasseril, Q.C. Liu, High-value SLM aerospace components: from design to manufacture. Adv. Mat. Res. **633**, 135–147 (2013). https://doi.org/10.4028/www.scientific.net/AMR.633.135
44. M. Yakout, M.A. Elbestawi, L. Wang, R. Muizelaar, Selective laser melting of soft magnetic alloys for automotive applications. *The 6th Joint Special Interest Group meeting between Euspen and ASPE: Advancing Precision in Additive Manufacturing* (2019) pp. 1–4. Accessed 30 Nov 2020. [Online]. Available: https://www.researchgate.net/publication/338717257_Selective_laser_melting_of_soft_magnetic_alloys_for_automotive_applications
45. D. Wang et al., Research on design optimization and manufacturing of coating pipes for automobile seal based on selective laser melting. J. Mater. Process. Technol. **273**, 116227 (2019). https://doi.org/10.1016/j.jmatprotec.2019.05.008
46. Y. Xiong et al., Fatigue behavior and osseointegration of porous Ti-6Al-4V scaffolds with dense core for dental application. Mater. Des. **195**, 108994 (2020). https://doi.org/10.1016/j.matdes.2020.108994
47. M. Antanasova, A. Kocjan, M. Hočevar, P. Jevnikar, Influence of surface airborne-particle abrasion and bonding agent application on porcelain bonding to titanium dental alloys fabricated by milling and by selective laser melting. J. Prosthet. Dent. **123**(3), 491–499 (2020). https://doi.org/10.1016/j.prosdent.2019.02.011
48. F. Bartolomeu, N. Dourado, F. Pereira, N. Alves, G. Miranda, F.S. Silva, Additive manufactured porous biomaterials targeting orthopedic implants: a suitable combination of mechanical, physical and topological properties. Mater. Sci. Eng. C **107**, 110342 (2020). https://doi.org/10.1016/j.msec.2019.110342
49. F.S.L. Bobbert et al., Additively manufactured metallic porous biomaterials based on minimal surfaces: a unique combination of topological, mechanical, and mass transport properties. Acta Biomater. **53**, 572–584 (2017). https://doi.org/10.1016/j.actbio.2017.02.024
50. M. Shafi, File:hip prostesis.png. *Wikimedia Commons*. Accessed 24 Sept 2024. [Online]. Available: https://commons.wikimedia.org/wiki/File:Hip_Prostesis.png
51. DepuySynthes, *DEPUY SMARTSET Gentamicin Instruction Leaflet.pdf*. (Johnson & Johnson, 2017)
52. DepuySynthes, *SMARTSET Bone Cements—Product Information*. (Johnson & Johnson Medical Limited, 2016). Accessed 18 July 2022. [Online]. Available: https://5.imimg.com/data5/MQ/RT/MY-62341139/depuy-smartset-1-bone-cement.pdf

Chapter 4
Development of Hip Implant

4.1 Introduction

This chapter outlines the procedures involved in this research to fulfil the research objectives of this study. It provides a comprehensive description of the phases involved in the development of designing and analysing the new hip prosthesis design. In general, the designing and analysis of the new hip implant design involved the usage of CATIA V5R21 and ANSYS 2020 R1 software to evaluate the results, preparation of jig and finally, compressive load testing for the hip prosthesis to obtain and compare the analysis with the simulation.

4.2 Design Selection

This part was done by using a Pugh chart to choose which of the existing hip prosthesis was the best choice to undergo a redesign. Table 3.1 provides a summary of the characteristics of existing hip prosthesis between Charnley and 3 M Capital, prior to the evaluation using the Pugh chart method. The datum for this research was the existing first generation of the Charnley hip prosthesis, and the criteria to be assessed were stability, load transfer, biocompatibility, wear and fracture resistances as in the Pugh chart in Table 4.1. Weight or scores were assigned to rank each existing prosthesis based on the criteria to aid in selecting the most suitable hip prosthesis for redesign purposes.

Table 4.1 Design selection through a Pugh chart

Design			Charnley		3 M Capital	
			Round back (2nd gen.)	Flanged (3rd gen.)	Round back	Flanged
Image[1]						
Criteria	Weight	Datum	Design 1	Design 2	Design 1	Design 2
Stability	3	0	+	+	+	0
Load transfer	2	0	+	+	+	−
Biocompatibility	2	0	+	+	−	+
Wear resistance	3	0	+	−	+	−
Fracture resistance	3	0	+	+	−	−

4.3 Finite Element Analysis (FEA)

FEA was carried out on the hip prosthesis as a stand-alone and when it was cemented inside a femur bone.

4.3.1 FEA of Hip Prosthesis

FEA was conducted on the hip prosthesis. The hip prosthesis was redesigned by using the CATIA V5R21 software and later imported to the ANSYS 2020 R1 software for this FEA process. Static structural analysis was performed to run the simulation. The femoral stem used was the Charnley stem and the material for the stem was titanium alloy. The analysed data were total deformations and equivalent (von Mises) stress.

The first draft of the modification was done as in Fig. 4.1a and later reviewed as in Fig. 4.1b. The first draft was reviewed, and the design was changed due to the analysis that did not really give any significant results to the hip prosthesis [1]. The main cause was that the stress shielding took place mainly at the neck of the hip prosthesis instead of at the stem of the implant. As a result, the design was modified so that a slot was added at the neck of the hip prosthesis instead of adding holes at the

---

[1] Used with permission of SLACK Incorporated, from Philosophies of stem designs in cemented total hip replacement, Nico Verdonschot, 28, 8 Suppl and 2005; permission conveyed through Copyright Clearance Center, Inc.

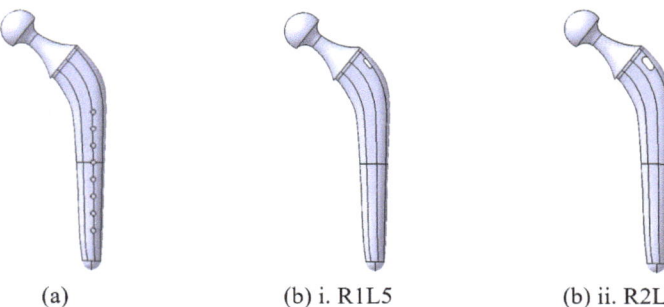

| (a) | (b) i. R1L5 | (b) ii. R2L5 |

Fig. 4.1 Charnley hip prosthesis: **a** First draft,[2] and **b** second draft of modification with slot, (i) of radius 1 mm and length 5 mm, and (ii) of radius 2 mm and length 5 mm

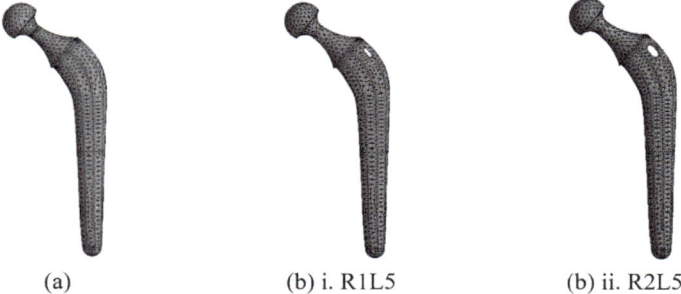

| (a) | (b) i. R1L5 | (b) ii. R2L5 |

Fig. 4.2 Mesh generated on the hip prosthesis: **a** Before,[3] and **b** after modifications of slot of 5 mm length with, (i) 1 mm radius, and (ii) of 2 mm radius

Table 4.2 No. of nodes and elements obtained after meshing process of the hip prosthesis

Hip prosthesis	Parts	No. of nodes	No. of elements
Existing		34,772	19,955
Modified	i. slot R1L5	37,254	21,313
	ii. slot R2L5	37,144	21,253

stem. A slot was added to promote porosity and for bone ingrowth and it is added at the neck to analyse the stress distribution and stress shielding on the hip prosthesis.

After selecting the materials and importing the designs, mesh convergence was done to determine the suitable element size. The method of meshing used was patch conforming tetrahedron method and 1.0 mm of the element size. Figure 4.2 shows the meshed designs while Table 4.2 portrays the number of nodes and elements produced through the meshing process.

---

[2] Reprinted from Hamizan et al. [1], with the permission of AIP Publishing.

[3] Reprinted from Hamizan et al. [1], with the permission of AIP Publishing.

Fig. 4.3 Boundary
conditions (head load and
fixed support) applied to the
hip prosthesis[4]

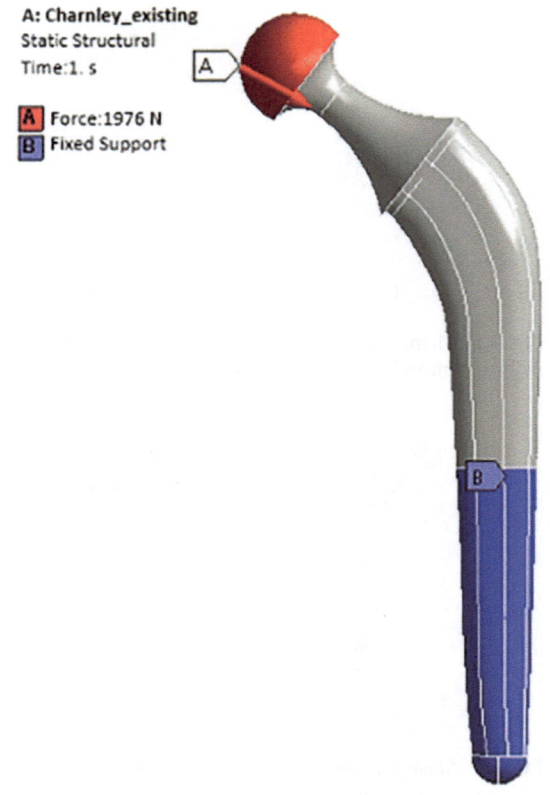

A: Charnley_existing
Static Structural
Time:1. s

[A] Force:1976 N
[B] Fixed Support

Once the meshing was generated, the boundary conditions were applied. This research only considered one scenario. It was the initial load case that corresponded to the first case which was the standing movement. The distal end of the hip implant was set as fixed support by fixing its x, y and z directions. The head of the stem was subjected to loads and boundary conditions as shown in Fig. 4.3. The head load value was 1,976 N [2, 3].

4.3.2   FEA of Cemented Hip Prosthesis Inside a Femur Bone

The femoral bone was assembled with the hip prosthesis and implanted into cement, which can be viewed in Fig. 4.4. A similar method of FEA was used for the bone assembly. The data that were analysed included the total deformations, equivalent (von Mises) stress and shear stress.

---

[4] Reprinted from Hamizan et al. [1], with the permission of AIP Publishing.

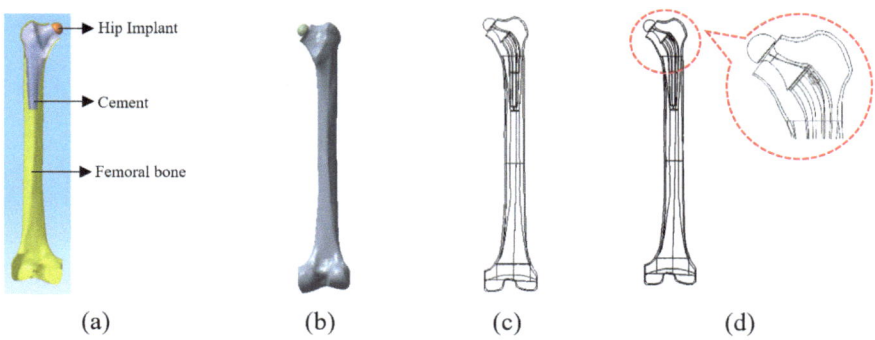

Fig. 4.4 Bone assembly: a Parts, b view in ANSYS software, c before, and d after modification with focus on the slot

After underwent material selection and importing of the designs, meshing was generated by auto-meshed for bone and cement while for hip prosthesis, it was done by using patch conforming tetrahedron method and 1.0 mm of the element size after mesh convergence. Figure 4.5 shows the meshed designs, while the view of cemented hip prostheses with different lengths of cement filings can be seen in Fig. 4.6. The cement fillings cut-off was obtained from previous research [4, 5]. Table 4.3 portrays the number of nodes and elements produced through the meshing process.

Boundary conditions were applied after the meshing was generated. Similarly, for this research, only one load case was considered. It was the initial load case that matched the first case, which was the standing movement. Loads and boundary conditions were applied to the stem as shown in Fig. 4.7. The distal end of the bone was secured in place by fixing its supports in the directions of the x, y and z axes. The head load value was 1,976 N [2, 6, 7], whereas the abductor load was 1,240 N [2, 3].

Fig. 4.5 Mesh generated on the bone assembly

Bone and PMMA:

Auto-meshed

Default element size

Hip prosthesis:

Patch conforming tetrahedron

1.0 mm element size

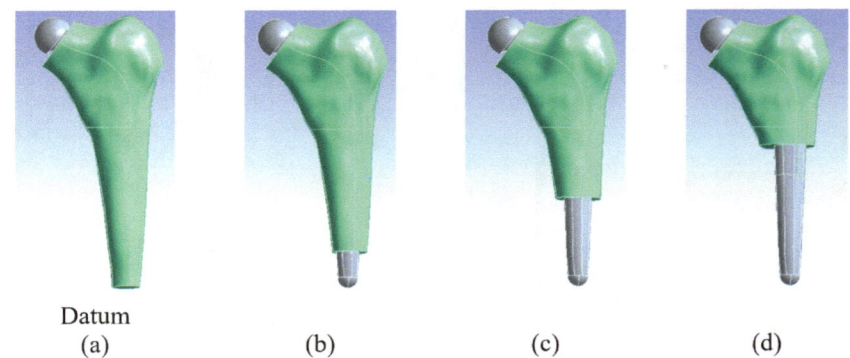

Datum
(a)   (b)   (c)   (d)

Fig. 4.6 Cement filing of the length of: **a** full or 156 mm, **b** 2/3 full or 134 mm, **c** 1/2 full or 104 mm, and **d** 1/3 full or 74 mm

Table 4.3 No. of nodes and elements on the bone assembly

Cement		Part	Existing		Modified R1L5		Modified R2L5	
			No. of					
			Nodes	Elements	Nodes	Elements	Nodes	Elements
(a)	Full cement/ Datum (156 mm)	Bone assembly	159,865	104,396	160,441	104,565	158,929	103,413
		Hip prosthesis	122,857	84,957	123,433	85,126	121,921	83,974
(b)	2/3 Full cement (134 mm)	Bone assembly	158,801	103,763	159,377	103,932	157,865	102,780
		Hip prosthesis	122,857	84,957	123,433	85,126	121,921	83,974
(c)	1/2 Full cement (104 mm)	Bone assembly	158,591	103,686	159,167	103,855	157,655	102,703
		Hip prosthesis	122,857	84,957	123,433	85,126	121,921	83,974
(d)	1/3 Full cement (74 mm)	Bone assembly	157,748	103,253	158,324	103,422	156,812	102,270
		Hip prosthesis	122,857	84,957	123,433	85,126	121,921	83,974

4.4 Preparation of Jig for Hip Implant

A customised jig was created to securely grip and exert pressure on the hip implant to carry out the compression testing process as illustrated in Fig. 4.8. The design of the jig follows the ISO 7206-4 Standard to comply with parts that were required for holding and applying pressure to the hip implant specimen during the compressive load testing.

Fig. 4.7 Boundary conditions (fixed support, head and abductor loads) applied to the bone assembly

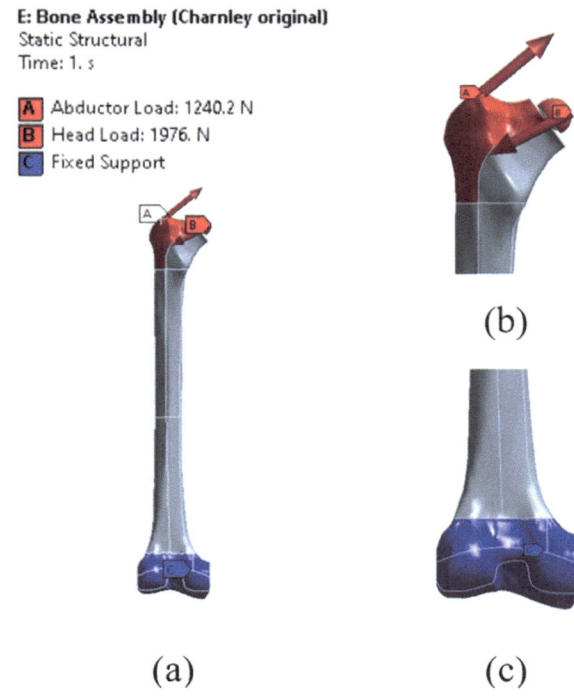

E: Bone Assembly (Charnley original)
Static Structural
Time: 1. s

A Abductor Load: 1240.2 N
B Head Load: 1976. N
C Fixed Support

(b)

(a)   (c)

Figure 4.8a shows the first draft of the custom-made jig that was made. Upon careful evaluation and review, the design was changed to optimise the utilisation of materials of the jig, which aimed to reduce waste and maximise the cost-effectiveness of the jig production process. Figure 4.8b illustrates the second draft of the jig design, whereby it had undergone design modification. The material chosen for the jig is stainless steel, known for its exceptional strength.

The customised jig was sent outsourced to be fabricated using stainless steel and later underwent a coating process to protect the jig from corrosion or damage during its usage, as shown in Figs. 4.9 and 4.10.

4.4.1 Fabrication of Hip Implant by Additive Manufacturing (AM) Technology

The fabrication of hip prosthesis was sent outsourced. The hip implant models in Fig. 4.11 were fabricated by using thefused deposition modelling (FDM) method utilising FlashForge 3D Printer. The material used for this model is polylactic acid (PLA).

First draft

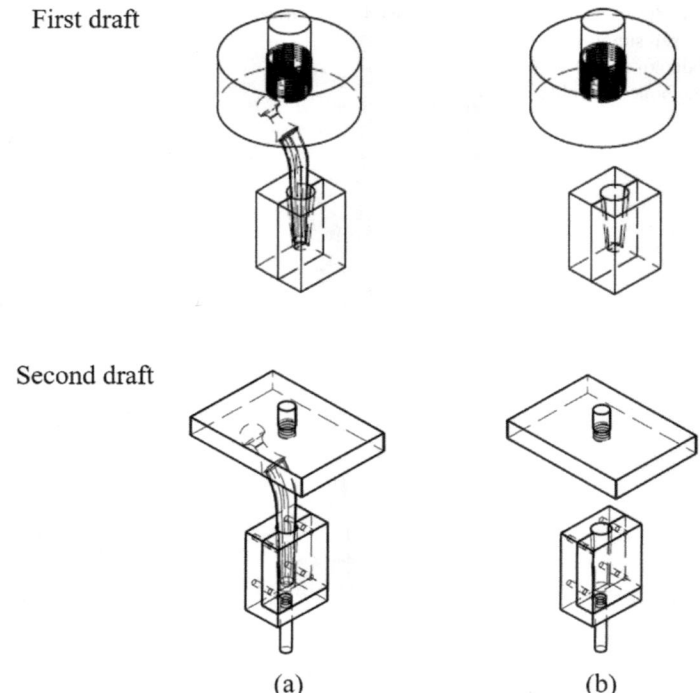

Second draft

(a)   (b)

Fig. 4.8 Illustration of the customised jig for first and second drafts, both at **a** with hip prosthesis, and **b** without hip prosthesis

Next, the hip prosthesis was fabricated by using SLM AM metal 3D printing technology. The process of fabricating the hip implant was made by using titanium alloy powder. Before the hip prosthesis underwent the testing procedure, some finishing was required to be done during the post-processing stage.

The fabrication of the hip implant was performed outsourced at our industrial collaborator, namely 3D Gens Sdn. Bhd., which is located at 18, Jalan Kerawang U8/108, Perindustrian Tekno Jelutong, Seksyen U8, Shah Alam, Selangor, Malaysia. The fabricated hip implant prototype was machine finished for its finishing process. One of the 3D printers available at 3D Gens is the Renishaw RenAM 500E Metal 3D Printer which is capable of the SLM process. Figure 4.12 shows the Renishaw RenAM 500E Metal 3D Printer machine [8] and Table 4.4 shows some printing parameters used by the 3D printer machine to print the hip implant.

Figure 4.13 shows the process of metal 3D printing the hip implant, during and right after printing, whereas Fig. 4.14 shows the hip implant before and after the finishing process.

(a)

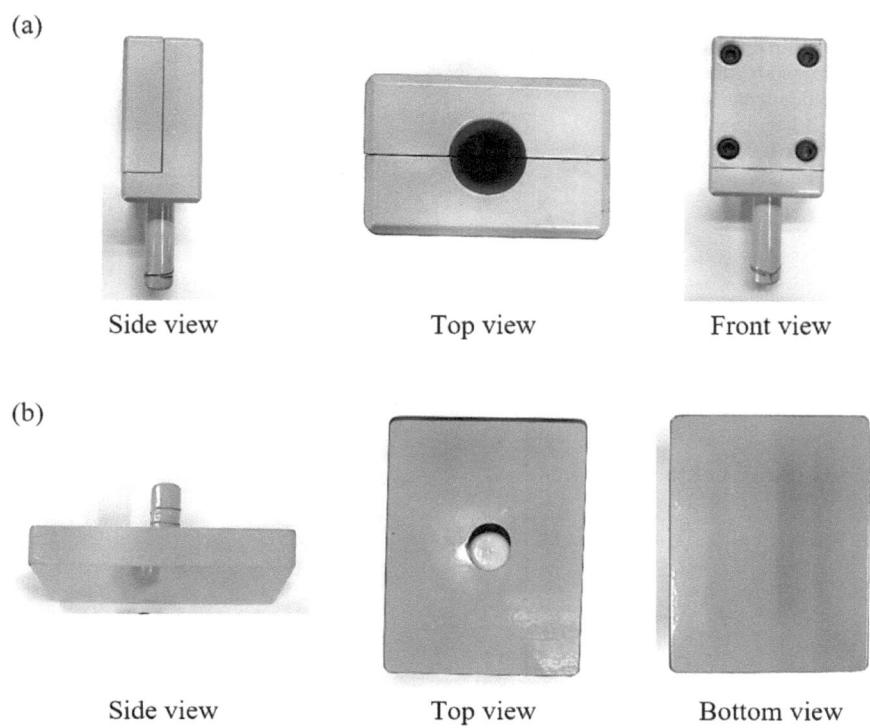

Side view   Top view   Front view

(b)

Side view   Top view   Bottom view

Fig. 4.9 Fabricated customised jig to: **a** Hold the hip prosthesis, and **b** when applying compressive load to the hip prosthesis during testing

Fig. 4.10 Overview of the fabricated jig and hip prosthesis for compressive load testing

Fig. 4.11 PLA hip implant model for hip prosthesis: **a** Before, and **b** after modification with slot of radius, (i) 1 mm and length 5 mm, and (ii) 2 mm and length 5 mm

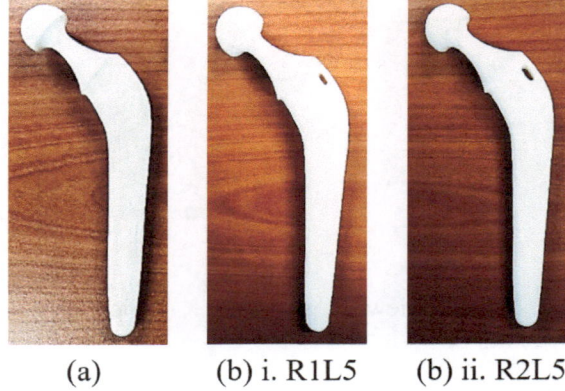

(a)   (b) i. R1L5   (b) ii. R2L5

Fig. 4.12 Renishaw RenAM 500E Metal 3D Printer [9][5]

---

[5] Reproduced with permission from Renishaw plc.

Table 4.4 Printing parameters of Renishaw RenAM 500E Metal 3D Printer that was used by 3D Gens Sdn. Bhd. to print the 3D hip implant

Feature	Parameter
Layer thickness	0.04 mm
Hatch distance	0.06 mm
Hatch increment angle	67°
Point distance	0.06 mm
Exposure time	0.05 s
Scanning speed	1.2 m/s
Laser power	200 W

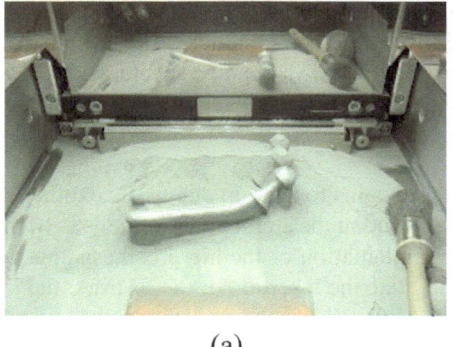

(a)

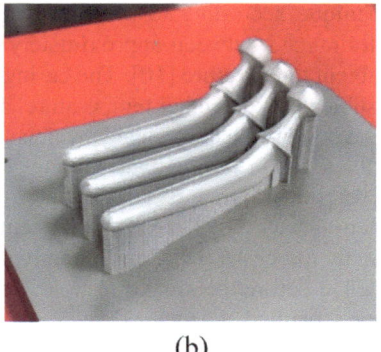

(b)

Fig. 4.13 Hip implant: **a** In the process of metal 3D printing, and **b** right after 3D printing

Fig. 4.14 Titanium alloy hip implant model for hip prosthesis after modification with slot of radius 2 mm and length 5 mm: **a** Before, and **b** after finishing process

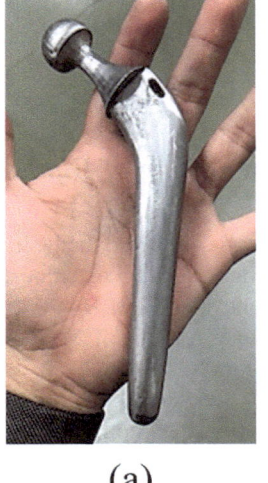

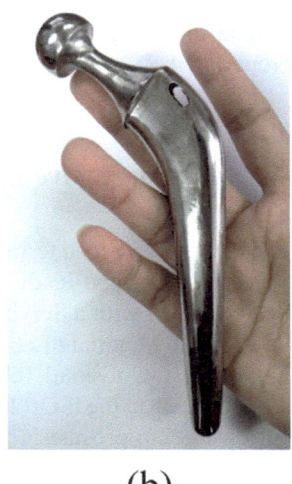

(a)

(b)

4.4.2   PMMA Preparation

For this compressive load testing, Gentamicin High Viscosity (GHV) PMMA provided by Johnson & Johnson's DePuy Synthes was used. To achieve the optimal or ideal PMMA processing times, certain requirements and conditions had to be carefully considered and followed. By adhering to the specified requirements and conditions, the compressive load testing aimed to attain the desired outcomes on the experimental test of the hip prosthesis.

There are several aspects that can influence the handling and processing times of PMMA. These factors include the ambient temperature, humidity and the mixing technique according to the manufacturer's instructions. The handling time typically ranged between approximately 13–15 min when conducted under consistent ambient temperature [10]. The curing process of PMMA can be accelerated when it is subjected to a higher temperature. The PMMA mixture is ready to be used once it has a dough-like consistency, which typically takes about 1–3 min, and is less sticky to the touch or during hand mixing.

As time progresses, the PMMA mixture begins to heat up due to the exothermic reaction caused by the release of chemical compounds after approximately 8–10 min into the process [10]. Consequently, the implantation of the hip prosthesis must be reached before this time threshold. To facilitate the insertion of the PMMA, the jigs were separated and the PMMA was carefully placed into the jig cavity. Subsequently, the hip prosthesis was implanted into the PMMA-filled jig cavity. The jigs were then reassembled and fastened together by using screws. Finally, the hip implant was manually secured, and the entire assembly was left undisturbed for a duration of 13–15 min, allowing ample time for the PMMA to fully cure and solidify.

The following Fig. 4.15 shows the visual representation of the procedure for preparing the PMMA.

4.4.3   Compressive Load Testing of Hip Implant

The compressive load testing was conducted on the hip prosthesis using a Shimadzu Servopulser E-type loading frame testing machine (Model: EHF-EV101K1-020-0A) equipped with a Servo Controller 4830. The testing machine is available at the Strength of Materials Laboratory, School of Mechanical Engineering, College of Engineering, Universiti Teknologi MARA.

According to the ISO 7206-4 Standard, the static compressive load test was carried out on the hip prosthesis samples. The compression tests were performed in two distinct conditions where one sample was tested with the presence of PMMA and the other without PMMA. By conducting both compressive load test experiments under both conditions, the analysis can be compared to the total deformations of the fabricated hip prosthesis and the influence of PMMA on the hip prosthesis. The analysis will then be compared and validated with previous research. Figure 4.16

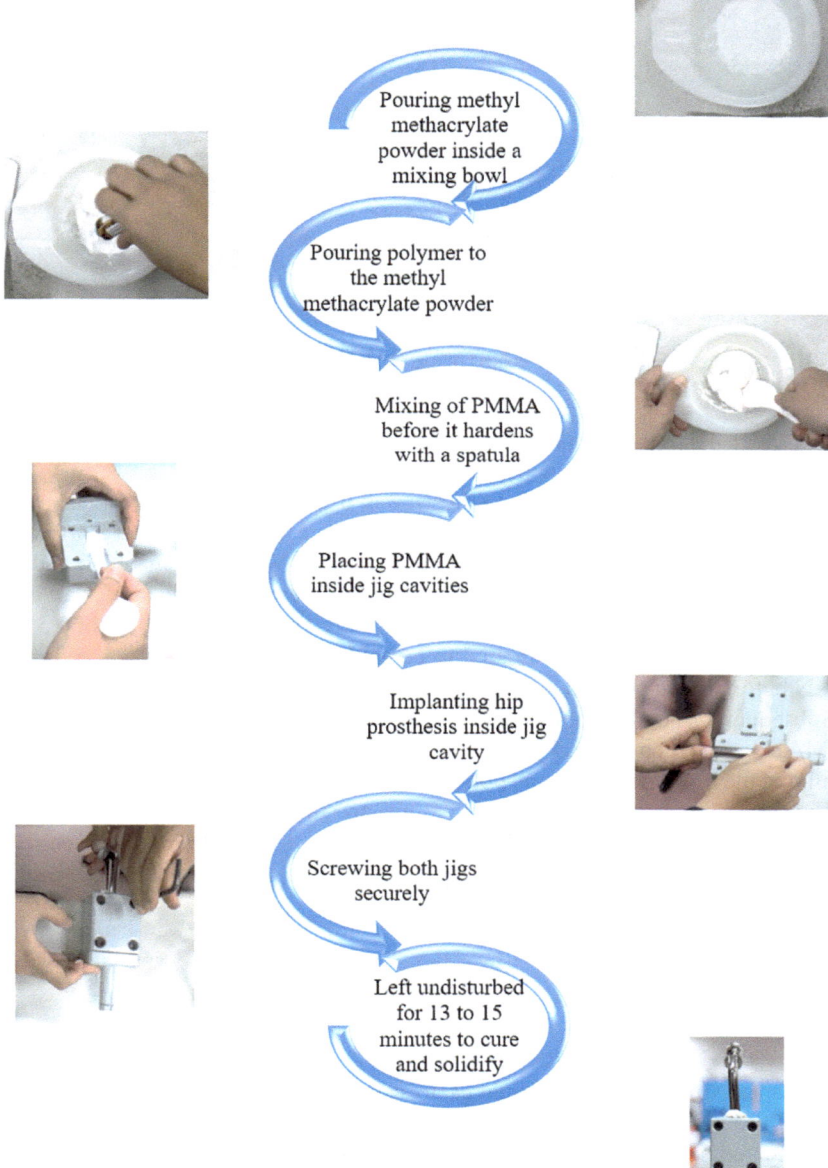

Fig. 4.15 Procedure to prepare PMMA

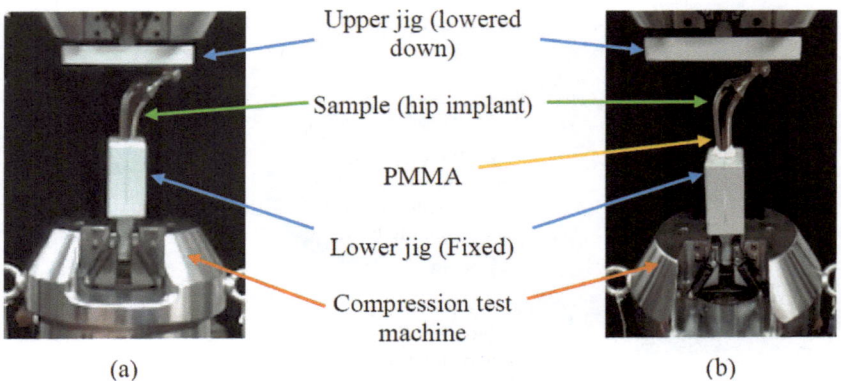

Upper jig (lowered down)

Sample (hip implant)

PMMA

Lower jig (Fixed)

Compression test machine

(a)   (b)

Fig. 4.16 Compressive load testing process of hip implant: **a** Without PMMA, and **b** with PMMA

provides a visual representation of the setup of the compressive load tests performed on the hip implants, both for with and without PMMA.

Firstly, a sample of the hip prosthesis was used for compressive load testing without the presence of PMMA, as shown in Fig. 4.16. The lower part of the customised jig was clamped securely onto the compression test machine and was left static or immobile during the experimental test. Meanwhile, the upper part of the jig, acting as the applied load, was gradually lowered at a controlled rate of 0.5 mm/min. The compressive load continued to be applied until it reached approximately 2.4 kN before being stopped manually.

Subsequently, another sample of the hip prosthesis was tested with the presence of PMMA as can be seen in Fig. 4.16. Prior to the test, the PMMA was carefully prepared by mixing the PMMA powder and poly liquid. The procedure for PMMA preparation can be referred to in Fig. 4.15. After the hip prosthesis was implanted within the PMMA in the jig cavity has cured, the compressive load testing was repeated by using the same compression test machine. The results of both experimental setups were compared and analysed.

4.5 Summary

This chapter summarises the methods to achieve the research objectives. These include the design selection on the hip implant, FEA on both stand-alone hip implant and cemented hip implant inside a femur bone, preparation of jig and PMMA, and compressive load testing of the hip implant. Chapter 5 shows the results and findings of the research, with graphical figures and tables to aid the analysis. The analysis will be discussed thoroughly and compared with previous research.

References

1. N.S. Hamizan et al., Design for additive manufacturing (DfAM) for hip prosthesis for 3D printing. AIP Conf. Proc. **2571**, 030005 (2023). https://doi.org/10.1063/5.0115923
2. S. Shuib, B.B. Sahari, W.S. Voon, M. Arumugam, Finite elemental analysis of outer and inner surfaces of the proximal half of an intact femur. Trends Biomater. Artif. Organs **26**(2), 103–106 (2012)
3. H.E.-D.F. El'Shiekh, Finite element simulation of hip joint replacement under static and dynamic loading. Doctor of Philosophy, Dublin City University (2002), http://doras.dcu.ie/17303/1/hussam_el-din_f._el-shie_20120702104151.pdf. Accessed 27 Jul 2020
4. M.A. Ahmad, N.N.M.E. Zulkifli, S. Shuib, S.H. Sulaiman, A.H. Abdullah, Finite element analysis of proximal cement fixation in total hip arthroplasty. Int. J. Technol. **11**(5), 1046 (2020). https://doi.org/10.14716/ijtech.v11i5.4318
5. M.I. Md Isa, S. Shuib, A. Z. Romli, A. Ahmed Shokri, I. Mohd Arrif, and N. S. Hamizan, "Finite Element Analysis (FEA) of the Different Cement Mixture for Total Hip Replacement", in 2021 IEEE National Biomedical Engineering Conference (NBEC) IEEE 2021 13 18 https://doi.org/10.1109/NBEC53282.2021.9618754
6. S. Shuib, N.F. Ismail, M.A. Yahya, A.A. Shokri, Analysis of an improved hybrid stem design for total hip replacement (THR). J. Mech. Eng. **5**(5), 205–215 (2018)
7. M. Viceconti, L. Bellingeri, L. Cristofolini, A. Toni, A comparative study on different methods of automatic mesh generation of human femurs. Med. Eng. Phys. **20**(1), 1–10 (1998). https://doi.org/10.1016/S1350-4533(97)00049-0
8. Data Sheet—RenAM 500E additive manufacturing system. Renishaw apply innovation, pp 1–4 (2019). https://www.renishaw.com/resourcecentre/en/details/Data-sheet-RenAM-500E-additive-manufacturing-system--113696?lang=English&srsltid=AfmBOorXvJcX6LD0bShiw3v2OWC_jfCbmJs5KHf61lL171r3QhphtN_j
9. RenAM 500E additive manufacturing system, in *Renishaw apply innovation*, pp. 1–6 (2019). https://tech-labs.pro/sites/default/files/Brochure_RenAM_500E_additive_manufacturing_system.pdf. Accessed 13 Dec 2020
10. DepuySynthes, DEPUY SMARTSET Gentamicin Instruction Leaflet.pdf (2017). Johnson & Johnson.

Chapter 5
Finite Element Analysis (FEA) of Modified Design of Hip Implant

5.1 Introduction

This chapter presents the results and findings of the research, compromises of design of the hip implant selected, FEA both on the hip implant and bone assembly with hip implant, and finally comparison between compressive load testing of the hip implant with and without the presence of PMMA. The analysis is discussed thoroughly with each data validation compared with previous research. Graphical figures and tables are used to support the data.

5.2 Design of Hip Implant Selected

Following the Pugh chart analysis in Table 4.1, the existing design of the hip prosthesis with the highest scores across the specified criteria was selected for further consideration. The scores from Pugh chart in Table 4.1 were calculated and viewed in Table 5.1.

Based on the Pugh chart method, design 1 of the Charnley hip prosthesis scored the highest through each criterion and was chosen to be redesigned as it is more stable, more biocompatible, has better load transfer, better wear and fracture resistance than the other three designs of Charnley and 3M Capital existing hip prosthesis.

After selecting the best design for the hip prosthesis, modification is done via CATIA V5 software. A slot for design modification was created by the usage of the Pocket tool with parameters of 5 mm length and two different radii: one with 1 mm and another with 2 mm, both starting at 5 mm from the femoral neck of the hip stem prosthesis, as shown in Fig. 5.1.

Table 5.1 Design selection through a Pugh chart

Design			Charnley		3M Capital	
			Roundback (2nd gen.)	Flanged (3rd gen.)	Roundback	Flanged
Image[1]						
Criteria	Weight	Datum	Design 1	Design 2	Design 1	Design 2
+			13	10	8	2
0			0	0	0	3
−			0	3	5	8
Net score			*13*	7	3	− 6

Bolditalic *13* is used to highlight the highest mark or the most favourable option in the Pugh chart, indicating the preferred choice based on the evaluation criteria

Fig. 5.1 Charnley hip prosthesis: **a** Before,[2] and **b** after modifications with slot, (i) of radius 1 mm and length 5 mm, and (ii) of 2 mm and length 5 mm

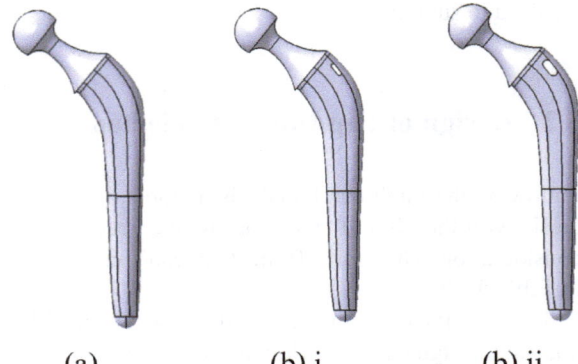

(a)   (b) i.   (b) ii.

5.3   Finite Element Analysis (FEA) on the Hip Implant

This section presented the overall FEA of total deformation, equivalent (von Mises) stress and shear stress.

---

[1] Used with permission of SLACK Incorporated, from Philosophies of stem designs in cemented total hip replacement, Nico Verdonschot, 28, 8 Suppl and 2005; permission conveyed through Copyright Clearance Center, Inc.

[2] Reprinted from Hamizan et al. [1], with the permission of AIP Publishing.

5.3.1   FEA of Hip Prosthesis

The overall analysis of total deformation and equivalent (von Mises) stress is shown in Fig. 5.2, respectively, for both the existing and modified design of the hip prosthesis, and the overall results are tabulated in Table 5.2.

The hip prosthesis with a 2 mm radius slot had higher maximum total deformation and lower maximum equivalent (von Mises) stress with 0.8 and 4.8% reduction from the existing design, respectively, than the hip prosthesis with a 1 mm radius slot.

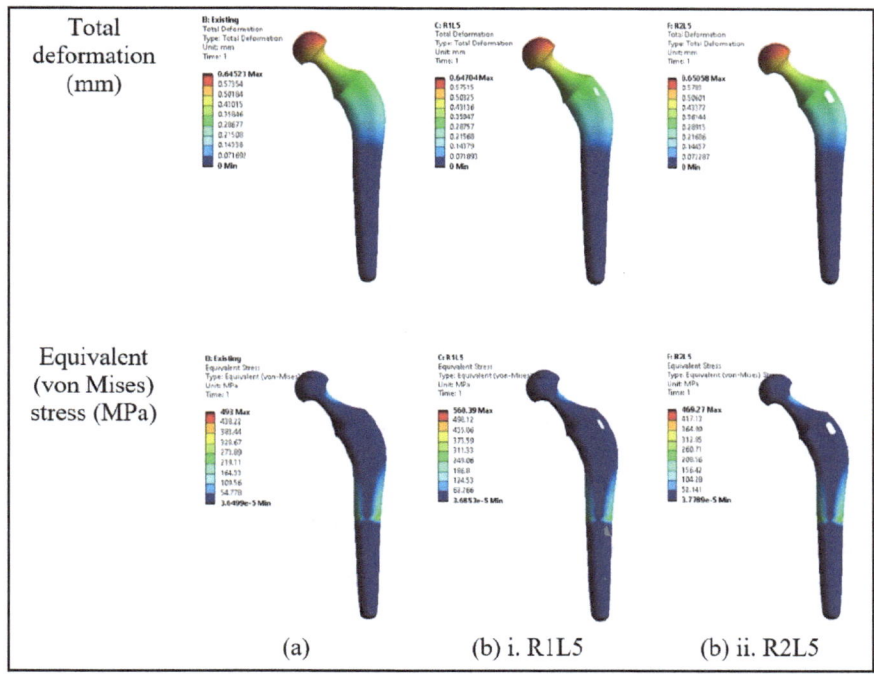

Fig. 5.2   Total deformation and equivalent (von Mises) stress for hip prosthesis: **a** Before and **b** after modifications with slot, i. of radius 1 mm and length 5 mm, and ii. of 2 mm and length 5 mm

Table 5.2   Overall analysis result of the hip prosthesis

Hip prosthesis		Maximum total deformation (mm)	Maximum equivalent (von Mises) stress (MPa)
Existing		0.6452	493.00
Modified	Slot R1L5	0.6470	560.39
		0.281%	13.67%
	Slot R2L5	0.6506	469.27
		0.829%	− 4.813%

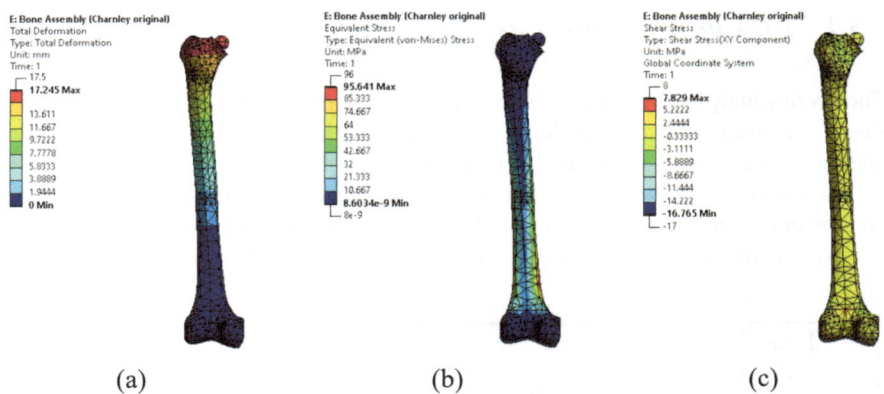

(a)   (b)   (c)

Fig. 5.3 FEA for bone assembly with cemented existing hip prosthesis inside a femur bone: **a** Total deformations (mm), **b** equivalent (von Mises) stress (MPa) and **c** shear stress (MPa)

5.3.2 FEA of Cemented Hip Implant Inside a Femur Bone (Bone Assembly)

The overall analysis of the bone assembly of cemented hip prosthesis on total deformation, equivalent (von Mises) stress and shear stress is shown in Fig. 5.3, respectively, with focused views on the hip prosthesis in Fig. 5.4. The analysed data were volume and mass of hip prosthesis, total deformations, equivalent (von Mises) stress and shear stress as tabulated in Tables 5.3 and 5.4, respectively, for bone assembly of the cemented hip prosthesis.

Table 5.3 shows the volume and mass of the hip prosthesis in the bone assembly. The hip prosthesis with a 2 mm radius slot had a lower volume and mass at about 1.25% reduction than the hip prosthesis with a 1 mm radius slot where it has about 0.48% reduction, both from the existing design. The 2 mm radius slot design achieved the AM of having a lesser volume and mass of a design.

Next, from Table 5.4, the length of bone cement can influence the analysis done via FEA. The lesser the length of the cement, the higher the percentage difference of total deformation, equivalent (von Mises) and shear stresses of the bone assembly. However, when reaching 1/3 or 74 mm of full cement, the analysis increases again. Therefore, the 1/2 or 104 mm of full cement has the most improved equivalent (von Mises) stress and shear stress amongst the other three cement mantle lengths like the previous findings [2]. This allows for minimising the stress shielding on the bone assembly with the least amount of cement mantle length that can be used.

In terms of the slot design modification, the hip prosthesis with a slot of 2 mm radius had a higher percentage difference of total deformation and equivalent (von Mises) stress but a lower or nearly similar percentage difference of shear stresses than that of the hip prosthesis with a slot of 1 mm radius. The modified Charnley hip prosthesis with a 2 mm slot design had higher total deformations, and lower

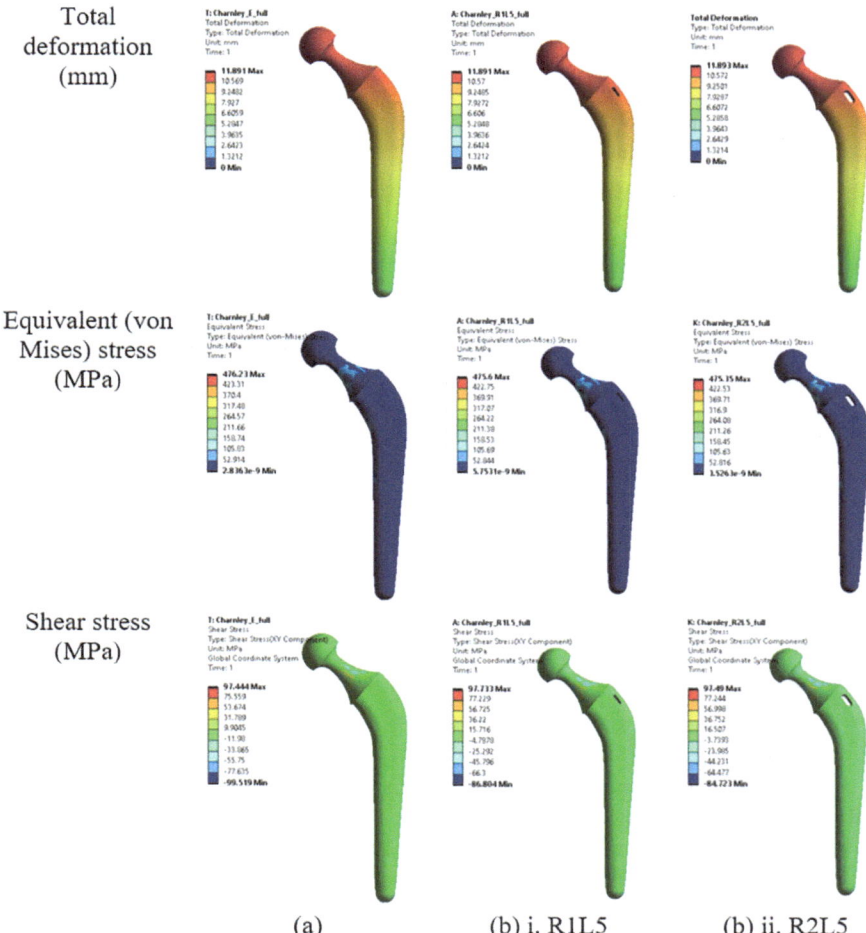

Fig. 5.4 FEA of total deformation, equivalent (von Mises) stress and shear stress for focused view of hip prosthesis inside the bone assembly: **a** Before and **b** after modifications with a slot, (i) of radius 1 mm and length 5 mm, and (ii) of 2 mm and length 5 mm

equivalent (von Mises) and shear stresses than the existing design and the 1 mm slot modification. The hip prosthesis with a 2 mm radius slot had a lower volume and mass and boasted a reduction of 1.25% than the hip prosthesis with a 1 mm radius slot and the existing design. Previous research by Arabnejad et al. [3] came out with porous structure hip prosthesis that can reduce stress shielding than fully solid hip prosthesis. These thus meet the objective of AM to minimise the weight (volume and mass) of the model with better efficiency, as opposed to the conventional method of production.

Table 5.3 Overall analysis result of the bone assembly by different height of cement-filled on volume and mass of the hip implant

Cement		Part	Existing	Modified R1L5	Modified R2L5
(a)	Full Cement/ Datum (156 mm)	Volume of hip implant (mm^3)	31,999	31,845	**31,598**
				− 0.481%	**− 1.253%**
		Mass of hip implant (kg)	0.14304	0.14235	**0.14124**
				− 0.482%	**− 1.258%**
(b)	**2/3 Full Cement (134 mm)**	**Volume of hip implant (mm^3)**	**31,999**	**31,845**	31,598
				− 0.481%	− 1.253%
		Mass of hip implant (kg)	**0.14304**	**0.14235**	0.14124
				− 0.482%	− 1.258%
(c)	1/2 Full Cement (104 mm)	Volume of hip implant (mm^3)	31,999	31,845	**31,598**
				− 0.481%	**− 1.253%**
		Mass of hip implant (kg)	0.14304	0.14235	**0.14124**
				− 0.482%	**− 1.258%**
(d)	1/3 Full Cement (74 mm)	Volume of hip implant (mm^3)	31,999	31,845	**31,598**
				− 0.481%	**− 1.253%**
		Mass of hip implant (kg)	0.14304	0.14235	**0.14124**
				− 0.482%	**− 1.258%**

Bold values is used to show that the modified R2L5 hip implant and 2/3 full cement (134 cm) were the preferred choice based on the evaluation criteria

Upon careful consideration of all the criteria including the cement mantle length, volume, mass and the slot design modification of the hip prosthesis, thus, the hip prosthesis with a 2 mm radius was chosen to be fabricated via AM 3D printing for the compressive load testing. This finding of hip prosthesis with a 2 mm slot design, particularly in terms of stress distribution and weight reduction, may contribute towards the advancement of prosthesis design through the AM techniques.

5.4 Compressive Load Testing for Hip Implant

Figure 5.5 shows the conducted compressive load testing done on a cemented hip prosthesis with a 2 mm radius and 5 mm length, whereas Fig. 5.6 shows the FEA done through ANSYS software to validate the analysis of total deformation of the modified hip prosthesis.

The comparison results for the simulation and compressive load testing experiment are presented in Table 5.5 and Fig. 5.7, with a maximum load given was 1.97 kN. The maximum deflection obtained at 1.97 kN was between 0.99 and 1.52 mm for all cases, which is in accordance with the specifications of ISO 7206-4 [4, 5]

Table 5.4 Overall analysis result of the bone assembly by different height of cement-filled on total deformations, equivalent (von Mises) stress and shear stress

Cement		Part	Existing	Modified R1L5	Modified R2L5
(a)	Full Cement/ Datum (156 mm)	Maximum total deformation (mm)	11.891	11.891	**11.893**
				0.000%	**0.017%**
		Maximum equivalent (von Mises) stress (MPa)	476.23	475.600	**475.350**
				− 0.132%	**−0.185%**
		Maximum Shear Stress (MPa)	97.444	97.733	**97.490**
				0.297%	**0.047%**
(b)	**2/3 Full Cement (134 mm)**	**Maximum total deformation (mm)**	**12.308**	**12.315**	12.313
				0.057%	0.041%
		Maximum equivalent (von Mises) stress (MPa)	**467.92**	**466.960**	466.670
				− 0.205%	− 0.267%
		Maximum shear stress (MPa)	**92.028**	**79.238**	79.123
				− 13.898%	− 14.023%
(c)	1/2 Full Cement (104 mm)	Maximum total deformation (mm)	12.241	12.247	**12.247**
				0.049%	**0.049%**
		Maximum equivalent (von Mises) stress (MPa)	447.41	446.480	**466.180**
				− 0.208%	**4.195%**
		Maximum shear stress (MPa)	91.384	78.785	**78.718**
				− 13.787%	**− 13.860%**
(d)	1/3 Full Cement (74 mm)	Maximum total deformation (mm)	13.305	13.312	**13.311**
				0.053%	**0.045%**
		Maximum equivalent (von Mises) stress (MPa)	500.32	498.510	**498.060**
				− 0.362%	**− 0.452%**
		Maximum shear stress (MPa)	100.42	84.582	**84.634**
				− 15.772%	**− 15.720%**

Bold values is used to show that the modified R2L5 hip implant and 2/3 full cement (134 cm) were the preferred choice based on the evaluation criteria

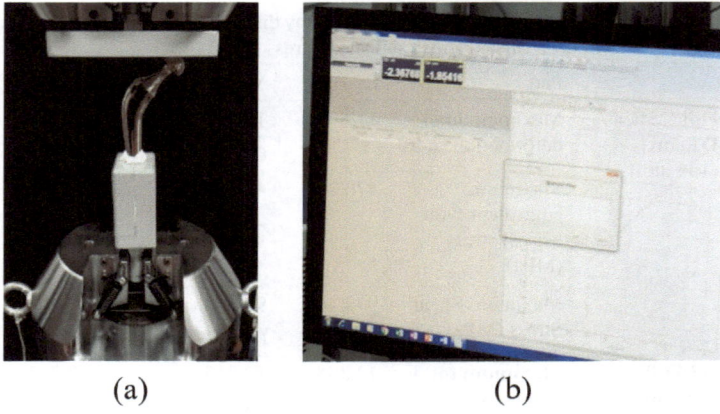

(a)   (b)

Fig. 5.5 Compressive load testing conducted: **a** Modified R2L5 hip prosthesis and PMMA and **b** data taken during testing on a PC

that indicate a maximum allowable 5 mm deflection for a hip implant embedded in a medium.

Although there were slight differences in the results, however, similar trendlines were produced through the comparison between the simulation or FEA method and the compressive load testing experiment results as shown in Fig. 5.7. Therefore, a good agreement can be achieved by this comparison method. There are several reasons that can influence the result readings.

The modified Charnley hip prosthesis design achieved a similar trendline via experiment and FEA methods. The specimens were able to withstand up to 2.5 kN load through both techniques, whereby given that the required endurance limit from the ISO 7206-4 for dynamic motion is 1.2–2.3 kN [4, 5]. The presence of PMMA makes it steeper and has lower total deformation than when it was absent at about 30% reduction during the experimental and 10% reduction as compared to the FEA.

From the line chart, the hip implant with PMMA has steeper lines than without the PMMA. For Sample 1 which was without PMMA, the hip implant was manually rested at one side of the jig, thus may cause the hip implant to move during the compression testing and creating an additional force from the jig. This indicated that it experienced a greater stiffness, thus may increase the stress shielding of the hip prosthesis, than when in the presence of PMMA.

Aside from that, from the comparison, the compression testing results both had much leaner lines than the simulation results with 27% without the presence of PMMA and 6% with its presence. This could happen because the boundary conditions of the hip implant when placed inside the jig are not totally rigid and could be due to PMMA preparation. The PMMA mixture is a subjective matter. For Sample 2 with PMMA, the temperature and humidity of the laboratory can easily affect the PMMA mixture. The PMMA was prepared in a little bit clumsily way since it was done for

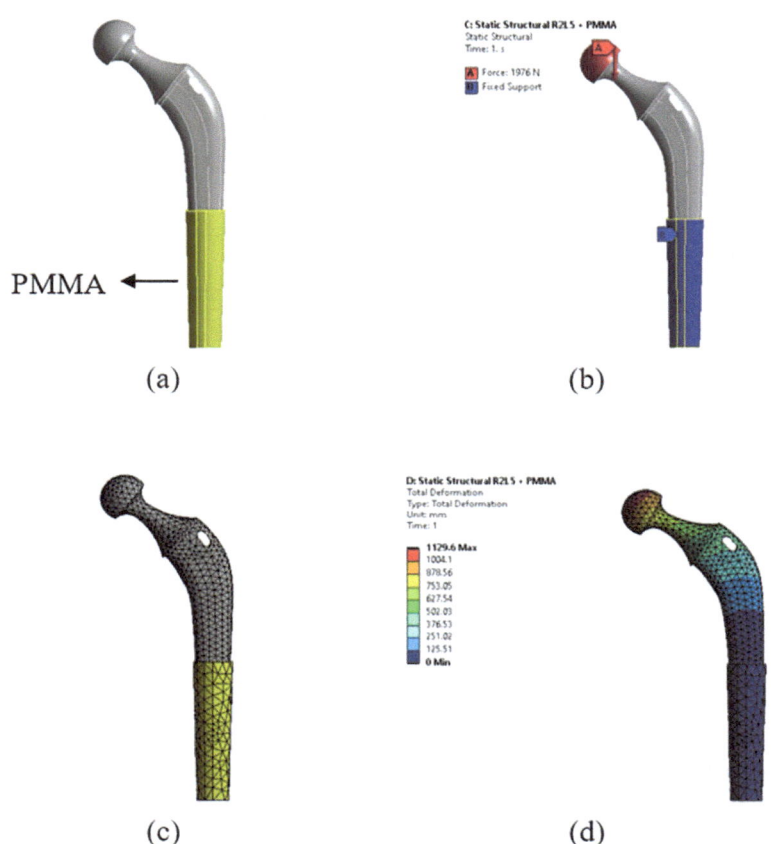

(a)   (b)

(c)   (d)

Fig. 5.6 FEA done: **a** Modified R2L5 hip prosthesis and PMMA, **b** loading and boundary conditions applied, **c** mesh generated and **d** total deformation of the hip prosthesis and PMMA

Table 5.5 Comparison of simulation and testing on total deformations of the hip prosthesis

	Hip implant without PMMA	Hip implant with PMMA	Percentage difference (%)
	Maximum total deformation (mm) at force 1.97 kN		
FEA	1.11	0.99	10.09
Compression test	1.52	1.06	29.97
Percentage difference (%)	26.93	6.19	

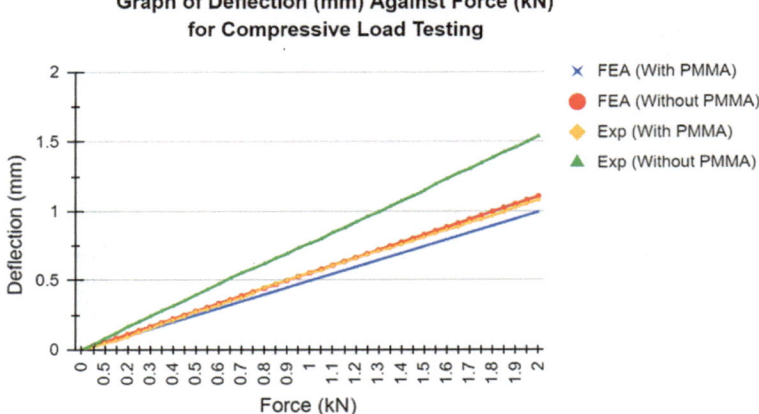

Fig. 5.7 Graph of deflection (mm) against force (kN) of FEA and compressive load testing of the hip prosthesis, with and without PMMA

the first time and not by a professional. The hip implant was then manually aligned to the jig.

Another possible reason for the higher percentage differences obtained between the comparison of the experimental and simulation could be due to the slice setting during the fabrication of the modified 3D printed hip prosthesis, which includes layer height, infill density, support structure, printing speed and temperature, that were beyond the scope of this study as the fabrication was done outsourced.

The compression testing analysis was similar to the analysis obtained by previous research by Čolić et al. [6, 7] and Suyitno et al. [7]. In their two research papers, Čolić et al. [6, 7] obtained that the static test had a 15% difference to the FEA result, twice that of this research. This is to be expected as they used different designs of hip prosthesis with slimmer femoral stem, big femoral head, had bigger slots and applied larger load force as they used the normal walking position.

Apart from that, Suyitno et al. [8] obtained 1.5 mm deformation at a 2.0 kN force load intersection. Their result is, however, slightly higher than the data obtained for this study as they used PMMA resin instead of PMMA cement. Although both types of PMMA can be used in real-life treatment, PMMA cement is however stiffer than the PMMA resin, thus validating that this research had higher total deformation than the previous research by Suyitno et al. [8].

5.5 Result Summary

First and foremost, the Pugh chart was used to identify and evaluate the properties and characteristics between existing Charnley and 3M Capital hip prosthesis designs. By weighing several factors in its evaluation, the best hip prosthesis design was chosen which was second generation Charnley hip prosthesis.

This research subsequently proceeded to develop a 3D model of a hip prosthesis using CAD modelling. The chosen Charnley hip prosthesis design was then further modified by adding a slot on the neck of the hip prosthesis and analysed the designed hip prosthesis by the usage of the FEA technique. The 1/2 or 104 mm of full cement exhibits the most improved analysis on the equivalent (von Mises) stress and shear stress regarding cement mantle length used during FEA of bone assembly with PMMA. Overall, the modified Charnley hip prosthesis with a 2 mm slot design was found to be the hip prosthesis that had optimum stress distribution, minimised stress shielding and weight.

Finally, compressive load testing was conducted on the fabricated hip prosthesis by designing and developing a suitable jig and the result of total deformation with FEA was compared, both with and without the presence of PMMA. The chosen 2 mm modified Charnley hip prosthesis was fabricated by utilising AM. Aside from that, customised stainless-steel jig and PMMA were prepared before underwent the compressive load test. For validation purposes, total deformation was analysed and compared between FEA and experimental test for both with and without PMMA., The experimental value and without PMMA were stiffer than FEA and with PMMA, respectively, from the comparison of analysis.

5.6 Summary

This chapter summarises the analysis of results and findings of the research with data validation subjected to compressive loading as compared to previous research. This included the design of the hip implant selected, FEA on both stand-alone hip implant and cemented hip implant inside a femur bone, by AM technology to fabricate the hip implant, and comparison of analysis from simulation and compression testing for the hip implant. Chapter 6 concludes on the research and recommends some future planning that can be done for this research.

References

1. N.S. Hamizan et al., Design for additive manufacturing (DfAM) for hip prosthesis for 3D printing. AIP Conf Proc **2571**, 030005 (2020). https://doi.org/10.1063/5.0115923

2. M.I. Md Isa, S. Shuib, A.Z. Romli, A. Ahmed Shokri, I. Mohd Arrif, N.S. Hamizan, Finite element analysis (fea) of the different cement mixture for total hip replacement, in *2021 IEEE National Biomedical Engineering Conference (NBEC)* (IEEE, 2021) pp. 13–18. https://doi.org/10.1109/NBEC53282.2021.9618754
3. S. Arabnejad, B. Johnston, M. Tanzer, D. Pasini, Fully porous 3D printed titanium femoral stem to reduce stress-shielding following total hip arthroplasty. J. Orthop. Res. **35**(8), 1774–1783 (2016). https://doi.org/10.1002/jor.23445
4. ISO, ISO 7206-4:2010—Implants for surgery—Partial and total hip joint prostheses—Part 4 Determination of endurance properties and performance of stemmed femoral components (International Organization for Standardization, 2010)
5. ISO, "ISO 7206-4:2010 AMD 1:2016—ISO 7206-4:2010—Implants for surgery—Partial and total hip joint prostheses—Part 4 Determination of endurance properties and performance of stemmed femoral components—Amendment 1 (International Organization for Standardization, 2016)
6. K. Čolić, A. Sedmak, A. Grbovic, U. Tatić, S. Sedmak, B. Đorđević, Finite element modeling of hip implant static loading. Proc. Eng. **149**, 257–262 (2016). https://doi.org/10.1016/j.proeng.2016.06.664
7. K. Čolić et al., Eksperimentalno i numeričko istraživanje mehaničkog ponašanja umjetnog kuka od legure titana [Experimental and numerical research of mechanical titanium alloy of hip implant]. Tehnicki Vjesnik [Technical Gazette] **24**(3), 709–713 (2017). https://doi.org/10.17559/TV-20160219132016
8. Suyitno, L. Sutomo, Static test and simulation of hip joint prosthesis, in *2017 7th International Annual Engineering Seminar (InAES)* (IEEE, 2017), pp. 1–4. https://doi.org/10.1109/INAES.2017.8068566

Chapter 6
Fundamental of Design and Development of Hip Implant

6.1 Introduction

This chapter serves as a comprehensive overview of the entire process of data collection and analysis undertaken throughout the course of this research. Furthermore, it also provides some suggestions and recommendations to strengthen this field for the advancement of this research.

6.2 Final Thoughts

In its entirety, this study has successfully addressed the three outlined main research objectives.

Firstly, research objective one aimed to evaluate the existing Charnley hip prosthesis design by identifying properties and characteristics of the existing hip prosthesis through the Pugh chart method. Through this approach of weighing several factors in evaluating its characteristics and properties, the best hip prosthesis design was chosen between existing Charnley and 3M Capital hip prosthesis, thus enabling to proceed in choosing the second generation Charnley hip prosthesis.

Following the completion of the initial objective, this research proceeded to develop and modify a 3D model of a hip prosthesis by using CAD modelling by adding a slot on the neck of the hip prosthesis, before further analysed. Analysis such as total deformation, equivalent (von Mises) stress and shear stress were obtained by employing the FEA technique, as a stand-alone prosthesis and cemented together with PMMA in bone assembly. Regarding cement mantle length, the 1/2 or 104 mm of full cement exhibited the most improved analysis on the equivalent (von Mises) stress and shear stress. Overall, this finding of the modified Charnley hip prosthesis with a 2 mm slot design, particularly in terms of optimum stress distribution, minimising

N. S. Hamizan et al., *Hip Prosthesis*,
SpringerBriefs in Applied Sciences and Technology,
https://doi.org/10.1007/978-981-96-1470-7_6

stress shielding and weight reduction, may contribute towards the advancement of prosthesis design through the AM techniques.

Finally, the final research objective aimed to conduct compressive load testing on the fabricated hip prosthesis by designing and developing a suitable jig and comparing the result of total deformation with FEA. Utilising AM by rapid prototyping, hip prosthesis was fabricated. Aside from that, customised jig and PMMA were prepared prior to the experimental test. Analysis of total deformation was compared between FEA and experimental test for both with and without PMMA, for validation purposes. From the comparison of analysis, experimental and without PMMA were stiffer than FEA and with PMMA, respectively.

In conclusion, the fulfilment of these objectives has significantly contributed to the understanding of this research topic and has significance towards the study that revolves around the development of a hip prosthesis that is redesigned, creating an optimum stress distribution and lower stress shielding to the modified hip implant, targeting to be fabricated by the usage of AM technology through rapid prototyping. Moving forward, future research could build upon these findings for future research or development with the intention of manufacturing it locally.

6.3   Recommendations

Nevertheless, this research is subject to several limitations and constraints. Since it is a biomechanical topic, it is essential to note that this research focuses only on the mechanical part of the research and utilises the latest technology of AM 3D printing. Consequently, the analysis might not consider the influence of the force of the muscle or tissue that is involved in real-life applications. Aside from that, the FEA is performed only under static structural motion. Therefore, the analysis could be affected if the simulation is done under dynamic motion and fatigue analysis. There are several opportunities to further improve the analysis. These include exploring different optimisation designs and techniques, the use of alternative materials for the hip prosthesis and incorporating a wider range of boundary conditions and loading types under various real-life scenarios that can influence the performance and durability of the hip prosthesis.

There was a challenge in removing the deposited PMMA that hardens inside the jig cavity during the fabrication process. As the jigs are intended to be used for future research, countermeasures should be considered for the removal of the PMMA from the jigs. One possible solution could involve redesigning the jigs with additional through holes and screws. This modification would facilitate the removal of both jigs from the PMMA. The idea is that both jigs will be forcefully pushed against one another during the removal of the additional screws of the through screw holes. Ultimately, while this research represents a commendable step towards advancing prosthetic design and analysis, it also serves as a call for continued innovation and refinement. By embracing these challenges and opportunities for improvement, we can push forward towards a more progressive and transformative changes in the orthopaedic and biomechanical industry.

Uncited References

1. DePuySynthes, Charnley Modular Hip System - Surgical Technique. Johnson & Johnson Medical Limited (2016). https://pdfslide.net/documents/charnley-modular-corail-8-depuy-syn thes-charnley-modular-hip-system-surgical-technique.html?page=1. Accessed 4 Nov 2020
2. C. Correa, A. Gil-Santos, J.A. Porro, M. Díaz, J.L. Ocaña, Eigenstrain simulation of residual stresses induced by laser shock processing in a Ti6Al4V hip replacement. Mater. Des. **79**, 106–114 (2015). https://doi.org/10.1016/j.matdes.2015.04.048

Index

© The Editor(s) (if applicable) and The Author(s), under exclusive license   73
to Springer Nature Singapore Pte Ltd. 2025
N. S. Hamizan et al., *Hip Prosthesis*,
SpringerBriefs in Applied Sciences and Technology,
https://doi.org/10.1007/978-981-96-1470-7